Ethical and Legal Dilemmas of Artificial Intelligence in Latin America

David Ramírez Plascencia
Rosa María Alonzo González
Editors

Ethical and Legal Dilemmas of Artificial Intelligence in Latin America

palgrave
macmillan

Editors
David Ramírez Plascencia
Centro Universitario de Guadalajara (CUGDL)
Universidad de Guadalajara (UdeG)
Guadalajara, Jalisco, Mexico

Rosa María Alonzo González
Facultad de Ciencias Humanas
Universidad Autónoma de Baja California
Mexicali, Baja California, Mexico

ISBN 978-3-031-86539-8 ISBN 978-3-031-86540-4 (eBook)
https://doi.org/10.1007/978-3-031-86540-4

Cover credit: Suriya Phosri / Alamy Stock Photo

This Palgrave Macmillan imprint is published by the registered company Springer Nature Switzerland AG.
The registered company address is: Gewerbestrasse 11, 6330 Cham, Switzerland

Contents

The Rise of Artificial Intelligence in Latin America 1
David Ramírez Plascencia and Rosa María Alonzo González
Introduction 1
AI Capabilities, Ethical Concerns, and Regulation 3
Aims and Scope 7
Organization of the Book 11
Final Remarks 14
References 15

Part I Considering AI in the Private Sphere 19

Artificial Intelligence, Copyright, and Media: An Analysis of Journalism in Latin America from an International Perspective 21
David Ramírez Plascencia
Introduction 21
Digital Journalism: From Paper to Bits 23
The End of the Print Press Model 24
Journalists' Labor Precarization 25
AI, Intellectual Property Rights, and Journalism in Latin America 27
Conclusions 31
References 33

From Epistemological Foundations over Futuristic Speculations to Fact-Based Concerns: Evolving Discussions on Artificial Intelligence in Argentina, Colombia, and Mexico (2014–2024) 41
Fátima Ávila-Acosta, Lucía Morales-Lizárraga, and Jan Nehring
Introduction 41
Background 43
Case Selection 44
Newspaper Selection 45
Methodology 46
Text Analysis with Large Language Models 47
Sentiment Analysis 48
Opinion Identification and Classification 48
Comparative Frequency Analysis 49
LLM Methods Evaluation 49
Results 50
General Trends and Distribution of AI-Related Opinion Columns 50
Sentiment Analysis 52
Opinion Classification 53
Comparative Frequency Analysis 55
Findings and Discussion 59
Growing AI Presence in Media and Refined Concerns 59
The Effects of the COVID-19 Pandemic and the Introduction of ChatGPT 59
National Trends and Localized Discussions 60
Limitations 61
Conclusions 62
References 63

The Algorithmic Tyranny in the Gig Economy: National Strategies and Policy Implications in the Latin American Context 67
Alisa Petroff
Introduction 67
Gig Economy in Latin America 70
Algorithms and Data Protection Regulations in Latin America 72
Argentina 73
Brazil 73

Chile 73
Colombia 74
Ecuador 75
Mexico 75
Uruguay 76
Conclusions 77
References 79

Mapping the Future of AI Regulation in Latin America: Civil Society Perspectives on Brazil's Pilot AI Regulatory Sandbox 85
Kenzo Soares Seto
Introduction 85
Brazil's Regulatory Landscape for AI 88
Multi-stakeholder Governance: The Controversial Voices of Society 89
The Rise of AI Regulatory Sandboxes 91
Forging a National Model: The Brazilian AI Regulatory Sandbox 92
Mapping Stakeholders in Brazil's AI Sandbox Consultation 93
Society Speaks: Brazilian Perspectives on the AI Regulatory Sandbox 94
What Should Be the Scope of the Brazilian Regulatory Sandbox(es) for AI? 95
The Battle for Algorithmic Transparency 96
Between Innovation and Compliance: Mechanisms of Regulatory Flexibility 97
Shedding Light on AI Governance and Multi-stakeholder Participation 98
Conclusion 99
References 102

Part II Challenges and Promises of AI in the Public Sector 105

Ethical and Legal Dilemmas of National AI Policies in Latin American Countries 107
Rosa María Alonzo González
Introduction 107
AI in Latin America: A North-South Technological Dilemma 108

AI Regional Policies 110
Incipient AI Policies in Latin America 110
An Actual Overview of AI Latin American Public Policies 112
Challenges, Risks, and Ethical Issues 119
Conclusions 121
References 122

The Use of Biometrics and AI for Border Control and Its Impact on Human Rights 125
Jezabel Pérez Yáñez and Natalia Brzezinski Ramírez
Introduction 125
Migration Crises, Mediated Refugees, and the Technologization of the Border Control 127
AI, Biometrics, and Migration: Legal and Ethical Concerns 129
Disruptive Technologies, Migrant Rights, and Civic Mobilization 133
Conclusions 135
References 137

Ethical and Educational Dilemmas of AI in Latin American Higher Education Institutions: Persistent Challenges and Inquiries 143
Jairo Alberto Galindo-Cuesta, Rubén Yáñez Reyna, and Paola Mercado Lozano
Introduction 143
Theoretical Framework 144
The Role of Ethics in GenAI 144
The Reformulation of Theoretical Definitions 146
Legislation on AI in Europe and Latin America 147
GenAI in the Context of Higher Education in Latin America 148
Methodology and Data Collection 150
Data Analysis 150
Participant Profile 150
Outcomes 151
Group 1: Privacy, Equity, and Ethical Actions (Fig. 1). 151
Group 2: Transparency, Consent, and Integrity (Fig. 2). 151
Group 3: Access, Development, and Other Concerns (Fig. 3). 152

Discussion 153
Ethical and Educational Dilemmas 156
Challenges and Questions 156
Conclusion 157
References 161

The Quest for a Responsible Use of AI in Latin America 165
Rosa María Alonzo González and David Ramírez Plascencia
Introduction 165
Volume Key Findings 168
The Quest for a Responsible Use of AI 173
Final Remarks 175
References 176

Index 179

Notes on Contributors

Rosa María Alonzo González holds a PhD in Social Sciences from the University of Colima. She is currently a member level 1 of the National System of Researchers (SNI) and is affiliated with Universidad de Baja California. Post-doctoral research for CONACYT at the Universidad de Guadalajara. She collaborates with Universidad de Guadalajara, and Universidad of Colima in Agorante, a research group in society and technology. Her works focus on the social uses of ICT and digital skills, digital production, and consumption, as well as communication and digital culture.

Fátima Ávila-Acosta is a doctoral researcher at the Berlin Graduate School of Social Sciences at Humboldt University in Berlin, Germany. Her doctoral research focuses on social discourses, attitudes, and perceptions toward violence against women in politics in Latin America. Her research interests also include participatory democracy, gender equality, and political representation. She integrates AI-based methodologies into her social research, particularly for gender-related analysis.

Natalia Brzezinski Ramírez studied her bachelor's in psychology at the University of Texas at Dallas and is currently studying her master's in business administration at the University of Louisiana Lafayette. Her academic interests include the impact of technology in society, particularly the spread of hate discourse online and migration. She served as an assistant at a migration refugee center in Dallas, TX from 2023 to 2034. Her latest publication is Ramírez, Brzezinski, and Alonzo. (2024) "Heroes and villains: The adverse context of medical staff during the COVID-19 pandemic in Latin America and Spain." ComHumanitas: Revista Científica de Comunicación, 15(2), Article 2.

Jairo Alberto Galindo-Cuesta is a doctoral researcher in Education at the National Distance Education University (UNED), focusing on ethics and didactics of artificial intelligence in higher education. He is a Full-Time Professor, Researcher, and University Lecturer with expertise in advising educational institutions on the design, analysis, and evaluation of educational experiences involving generative artificial intelligence and intelligent agents. Specialist in Artificial Intelligence for Content Generation (IEBS), E-Learning Project Management (UOC), and Virtual Learning Environments (OEI), he holds a Master's in Hispano-American Linguistics (Caro y Cuervo). His work centers on integrating pedagogy and technology, with an emphasis on ethical and innovative approaches to the implementation of AI in educational contexts.

Paola Mercado Lozano holds a Master in Learning Technologies and is Professor and tenured faculty member at the University of Guadalajara and is Coordinator of The Master in Learning Management in Virtual Environments. Her research interests focus on the study of the establishment of digital ecosystems of knowledge on virtual education. Her latest publication is Yáñez & Mercado (2023). Análisis de una experiencia de uso de herramientas de Moodle desde Google Classroom a través del uso de un esquema de autenticación unificada.

Lucía Morales-Lizárraga is a PhD candidate in the Doctoral Program of Cultural Management at the University of Guadalajara, Mexico. Her doctoral research explores othering discourses within the radical right in Germany, focusing on their role in political polarization. She holds an MA in Roads to Democracy(ies) from the University of Siegen, Germany, and her broader research interests include identity processes, emotional dynamics in polarization in Social Media, and conspiracy theories. Alongside her academic work, she is a consultant specializing in risk and data management.

Jan Nehring holds a PhD in Artificial Intelligence from the Technical University of Berlin. He worked as a researcher at the Speech and Language Technologies Group at the German Research Center for Artificial Intelligence (DFKI) in Berlin, Germany, and is currently working as a development consultant for the Centre for the Fourth Industrial Revolution in Rwanda. He is a computer scientist by training, and his research interests include Natural Language Processing, dialog systems, and large language models.

Jezabel Pérez Yáñez is a candidate for the bachelor's degree in forensic sciences at the University of Guadalajara. She has experience in the collection and evaluation of data from video surveillance systems and security technology monitoring applied to prevention and improvement of public security strategies. She has participated as a speaker at the Autumn Colloquium of BINACOM of the Autonomous University of Baja California 2024 with the theme research "TikTok and the American Dream. Let's talk about how to migrate," as well as at the International Congress of the XXIX Summer of Pacific Scientific and Technological Research 2024 with the research topic "The Use of TikTok Among Migrants as a Strategic Tool for Your Journey to North America."

Alisa Petroff is a Collaborator Researcher at Universitat Oberta de Catalunya, Internet Interdisciplinary Institute, and CER-Migracions (UAB), apetroff@uoc.edu, Parc Mediterrani de la Tecnologia (Edifici B3) Av. Carl Friedrich Gauss, 5 08860 Castelldefels (Barcelona). Her latest publications include: Petroff, A., & Fierro, J. (2023). Crowd work in STEM-related fields: A window of opportunity from a gender perspective? *Sociology Compass*, e13058. https://doi.org/10.1111/soc4.13058; Petroff, A. (2022). Crowdsourcing in ICT Fields: The Link Between Mobility and Digital Labor Markets in Highly Skilled Tasks. In: Baikady R., Sajid S., Nadesan V., Przeperski J., Islam M.R., & Gao J. (eds).

David Ramírez Plascencia is Professor at the University of Guadalajara, specializing in the study of information law and digital policies. His recent works include Dalton, D., & Ramírez Plascencia, D. (Eds.). (2023). Imagining Latinidad: Digital Diasporas and Public Engagement Among Latin American Migrants, and The Politics of Technology in Latin America (Volume 1): Data Protection, Homeland Security and the Labor Market and The Politics of Technology in Latin America (Volume 2): Digital Media, Daily Life and Public Engagement (Routledge, 2020). He has done fieldwork in diverse cities in Spain. His research focuses on the study of the possibilities and dilemmas of information technologies in developing countries. He is interested in digital activism, information control, and social media use in Latin America.

Kenzo Soares Seto holds a Ph.D. in Communication from the Federal University of Rio de Janeiro (UFRJ) and is currently beginning his post-doctoral studies at Yale Law School. He is a researcher associated with Digilabour, a former adjunct professor at UFRJ's School of Communication, and a founder and former member of NETLAB-UFRJ. His research focuses on the political economy and governance of AI in Latin America, as well as the socio-technical imaginaries of Latin American tech workers. Kenzo also has over a decade of experience in public policy, serving as a parliamentary advisor across municipal, state, and federal legislative bodies in Brazil.

Rubén Yáñez Reyna is currently a researcher at the Universidad de Guadalajara. Yáñez Reyna holds a master's in public services management in Virtual Environments. Yáñez Reyna has worked as software developer in diverse public institution projects such as México Conectado, Jalisco Government, and UDGVirtual. His research focuses on the application of technologies for learning and the management of virtual environments and learning processes. His latest publication is Yáñez & Mercado (2023). Análisis de una experiencia de uso de herramientas de Moodle desde Google Classroom a través del uso de un esquema de autenticación unificada.

List of Figures

From Epistemological Foundations over Futuristic Speculations to Fact-Based Concerns: Evolving Discussions on Artificial Intelligence in Argentina, Colombia, and Mexico (2014–2024)

Fig. 1	Number of articles per year	51
Fig. 2	Number of articles per year and country	51
Fig. 3	Yearly distribution of sentiments	52
Fig. 4	Distribution of sentiments per country	53
Fig. 5	Distribution of sentiments per year and per country	53
Fig. 6	Yearly distribution of opinions	54
Fig. 7	Distribution of opinions per country	54
Fig. 8	Distribution of opinions per year and per country	55
Fig. 9	Aggregated comparative frequency analysis	56
Fig. 10	Argentina: comparative frequency analysis	56
Fig. 11	Colombia: comparative frequency analysis	57
Fig. 12	Mexico: comparative frequency analysis	57

Ethical and Educational Dilemmas of AI in Latin American Higher Education Institutions: Persistent Challenges and Inquiries

Fig. 1	Group 1. Privacy, equity, and ethical actions	152
Fig. 2	Group 2. Transparency, consent, and integrity	153
Fig. 3	Group 3. Access, development, and other concerns	154
Fig. 4	Challenges and questions. Source. Authors	160

List of Tables

From Epistemological Foundations over Futuristic Speculations to Fact-Based Concerns: Evolving Discussions on Artificial Intelligence in Argentina, Colombia, and Mexico (2014–2024)

Table 1 Distribution of articles by country and newspaper 47
Table 2 Evaluation metrics for sentiment analysis and opinion classification 50
Table 3 Aggregated comparative frequency analysis per year 58

Ethical and Educational Dilemmas of AI in Latin American Higher Education Institutions: Persistent Challenges and Inquiries

Table 1 Key ethical and educational dilemmas 158
Table 2 Strategic recommendations 159

The Rise of Artificial Intelligence in Latin America

David Ramírez Plascencia
and Rosa María Alonzo González

Introduction

Although there is not a unique definition of Artificial Intelligence, it is commonly accepted that when we are referring to this term, we are talking about the technological developments that enable machines to imitate various complex human skills (Sheikh et al., 2023). These skills could embrace actions such as processing information and taking independent decisions without external intervention. The first attempts to theorize

D. Ramírez Plascencia (✉)
Centro Universitario de Guadalajara (CUGDL),
Universidad de Guadalajara (UdeG), Guadalajara, Jalisco, Mexico
e-mail: davidram@udgvirtual.udg.mx

R. M. Alonzo González
Facultad de Ciencias Humanas, Universidad Autónoma de Baja California,
Mexicali, Baja California, Mexico
e-mail: rosa.alonzo@uabc.edu.mx

D. Ramírez Plascencia, R. M. Alonzo González (eds.), *Ethical and Legal Dilemmas of Artificial Intelligence in Latin America*,
https://doi.org/10.1007/978-3-031-86540-4_1

about AI could be traced to the mid-twentieth century, more specifically in 1950, when some scholars, Alan Turing among them, started discussing how computers could be empowered to achieve human thinking processes (Agrawal et al., 2018). More than seven decades later, due to the popularization of chatbots such as ChatGPT (OpenAI), we are experiencing a growing interest in artificial intelligence at a global level. This enthusiasm, to some extent, is justified. The remarkable capacities of AI applications not only focus on maintaining conversations, but people can also ask for advice about buying a new computer or translating a chunk of information. They can even generate human-like thinking works such as academic essays and legal plaintiffs or create pieces of art such as poems or paintings.

However, this social trend about AI appeared in an intricate context shaped by political polarization, military conflicts in Europe and the Middle East, global migrations crises and a rising distrust toward economic aperture. The popularity of applications based on artificial intelligence could be seen as a tendency to search accuracy and reliability in an age characterized by the proliferation of disinformation and suspicion toward traditional institutions and media. As people stop trusting in big media corporations and old-style political parties, they are searching for a more direct approach to information that complies with their personal interest, and in some ways, their ideology. The dominant technological corporations are now trying to consolidate themselves in a digital-content market that is changing into the on-demand stream service model, where movies and music are "suggested" -some could argue that the content is "imposed"- and distributed to the user by artificial intelligence.

In September 2017, the Russian president Vladimir Putin stated that "Whoever leads in AI will rule the world" (RT International, 2017). Eight years after this declaration, the world is witnessing the irruption of a new "Cold War" shaped by digital nationalism, nearshoring, and economic protectionism. China has surged as a new global superpower not just in the economic or militarist sector, but also in the technological one. The United States, China, and the European Union are competing in a technological race not only about developing AI systems but also about controlling the infrastructure that supports it. Actions, such as Biden's administration decisions, now in the hands of Trump, to limit the exports of AI chips and chip-making tools to the Asian country or the enforcement of selling TikTok, a very popular social media app, owned by the Beijing-based ByteDance corporation to a non-Chinese company, are just responses to this rising distrust among two countries who want to control the future of AI and, therefore, the distribution of digital content across the globe.

AI Capabilities, Ethical Concerns, and Regulation

The proliferation of AI applications in economic, ludic, medical and military sectors, among others, has triggered not only hopes about their latent advantages, but also growing apprehensions and debates regarding some potential abuses from governments and corporations alike: such as the invasion of users' private sphere, illegal public surveillance of user's activity, concerns about trusting the managing of sensible systems, as the case of airports, sanitary and military facilities under the supervision of AI, and the legal and ethical issues of using automatized weapons in battlefields.

Considering the impact of AI on the environment, it is possible to find two countered edges. AI can, for example, assist public agencies and rural communities in their journey to protect ecosystems from illegal logging, mining, and furtive hunting. Tools such as drones and complex systems that process huge masses of data can monitor and alert about potential risks to natural protected areas (Dauvergne, 2021). On the other hand, the operation of AI requires huge energy. Digital hubs are high-power consumers, and their cooling systems need vast amounts of water. In the education sector, AI-based learning technologies provide promising solutions to support educational inclusion (du Boulay et al., 2023). This is the case of intelligent tutoring systems (ITSs). Adaptive instructional systems can improve learning outcomes and instructional traditional methods based on brick-and-mortar classrooms by personalizing the education experience and designing specific strategies for every student. Other ideas focus on providing "A pedagogical agent (PA)" a virtual character that assists, promotes, encourages and cheers on students to improve their performance (Chen, 2023). But there are also criticism regarding the use of AI in the classroom. Some people want to avoid their use, while others prefer a more flexible approach that permits their employment but with some limitations and rules, that students declare, for example, if she/he used AI to write an essay and in which sections of the essay or percentages were AI-collaborated.

Other debates focus on the introduction of AI in the industry; a very sensitive issue in Latin America, whose labor market depends substantially on the manufacturing of goods. A sector where the introduction of AI could eventually replace hundreds of non-qualified employers as it has happened across the years with the car-making industry in the region (Brambilla et al., 2023). However, recent studies about the impact of AI in the labor market are enthusiastic about how the uses of these "disruptive" developments can increase productivity by having AI models bring out the best labor practices, helping new workers to become more competent in less amount of time (Brynjolfsson et al., 2023). There are other applications that could eventually replace humans, such as chatbots that will change ecommerce and client assistance services by providing 24-hour information. Chatbots are moving into other areas such as the government that is using automated accounts on social media and WhatsApp to deliver public information and assist citizens with some procedures. AI will eventually bring benefits in the health area as well: supporting scientific tests to develop vaccines or improve medical procedures, helping doctors to make better diagnoses and establish more accurate treatments, using bots to guide patients who want to know more about their symptoms, and providing emotional support to patients during their treatment.

In the military sector, there have already been intense discussions, particularly considering the role that machines could play in the decision-making processes at battlefields. Countries such as the USA and China have invested great efforts on developing AI systems to improve homeland security. Software capable of analyzing huge amounts of data, making decisions, and pondering the risks in each situation (Haney, 2020). Recent conflicts in Nagorno-Karabakh and the actual Russo-Ukrainian War illustrate well the strategic role of drones, not just for surveillance purposes but also as weapons. During this conflict, the production of drones has cheapened and become more precise. Ukraine has been successful transforming domestic drones into high-precision weapons. They have even started improving UAV with basic AI systems that help the drone to avoid defenses and attack key infrastructure objectives near Moscow (The Economist, 2024a, 2024b). However, there are concerns about the AI capacity of discriminating between an unarmed civilian and a soldier (Sharkey & Sharkey, 2021) and ethical questions about civil and criminal liability if an autonomous weapon system attacks civil infrastructures such as a hospital or a school (Liao, 2020).

Amid this vibrant global competition to expand the potentialities of AI, there have been exploding cases that clash with the traditional legal and ethical boundaries. AI, as other communication technology such as the Internet, reflects the class, genre, and race inequalities present in the society that develops and "feeds" it. In recent years, there have been critics that point out how many AI apps deliver outcomes that could be considered racist or that provide a partial vision of facts that echoes an "imperialist rhetorics of colonization and conquest" (Katz, 2020, p. 07). Some academics stand that sometimes applications base their outcomes on race and gender bias, preferring men over women and white people over other races, inferring professions just by a facial photograph (Hundt et al., 2022). For that reason, they assert that it is important not to leave aside underrepresented groups, particularly in the Global South, and encourage them to create and disperse their own discourse on the Internet and social media, and to be critical, as well, about how AI replicates the traditional contradictions of our human world (REPEF, 2023). Other controversies in the political sphere relate with the use of AI bots in electoral campaigns (Panditharatne & Giansiracusa, 2023). Bots are becoming the more effective allies to political parties by using generative AI to create false multimedia elements that impact social media users and shift their political preferences. In fact, AI generative has been the subject of criticism because of its capacity to create fake content that could deceive people. It could generate images, videos and audios supporting or attacking public figures, falsely portraying them committing illegal acts. By now, there are other polemic cases related with the potentialities of artificial intelligence, such as the generation of pornographic images from material taken from the Internet. This not only invades the private sphere of individuals, but the exploitation of their images constitutes a fragrant attack to their dignity (Murphy Kelly, 2024).

Debates associated with the misuses of AI have echoed in the media across the years. Media, particularly newspapers, has been an important disseminator of information about novel technological innovations. Most of what the common people, those who are not experts in computer science, know about the potentialities, barriers, and risks of artificial intelligence comes from the media. The process of media reception among the journalists and the audience is not neutral but framed. Research about media coverage in the USA provides evidence on how artificial intelligence is usually portrayed with both contradicted visions: (a) optimist and (b) pessimist. There are positive visions about its potential impact on education, work, health, transportation, entertainment, and so on. But other news highlight latent risks such as the loss of control of AI developments, layoffs because of automation, and the lack of ethical and legal frameworks that put under the public scrutiny the operation of automated weapons, etc. (Fast & Horvitz, 2017). UK media coverage highlights the great impact of AI developments to solve traditional social and economic problems but as Brennen et al. (2018, p. 04) state "rarely do journalists or commentators question if new technologies are the best solutions to the myriad of problems." In Latin America, the media coverage presents this dichotomy as well, portraying AI as a source of fears about the negative impacts in the labor market but there are also hopes about the great impact to improve people's lives. AI is sometimes associated with the literary figure of Frankenstein, as a technology that may surpass ethical boundaries and could turn against humans (Ramírez Plascencia, 2020).

These fears and hopes caused by the irruption of AI applications in peoples' daily lives have translated into ethical and legal discussions such as the limits in where these applications are able to perform, the legal liability of the AI corporations, users' legal protections and the ethical and legal frameworks that must guide actual and future AI developments. These issues remain central since there are cases where, for example, the AI applications provide benefits and advantages for their users but at a great cost, such as their privacy and the integrity of their personal data (Floridi, 2021). Along with the protection of user's private data from potential misuses and economic exploitation, there are other novel key questions that have arisen as people incorporate AI to support both daily life activities and public systems and security issues as well. Questions such as what could be the economic, ethical and security consequences of replacing human decision-making with AI? (DiMatteo et al., 2022). It is imperative that this substitution process should be made with transparency and under a set of clear and compressive regulations.

In these settings, it is crucial to discuss the liability of crimes, accidents, and illegal actions committed by the AI application or the users. In the case of accidents, technological corporations have traditionally focused on blaming the humans above the machine, "counter-intuitive focus on human responsibility even while human action is increasingly replaced by automation" (Elish & Hwang, 2015). This has been the case of commercial aircraft accidents. But what will be the legal criteria regarding autonomous cars such as the drive-less taxis that circulate in some cities in the United States? Who will be liable for an accident? The authority that approves the operation, the AI corporation that develops the system that controls the vehicle, or the company that provides the service? Regarding the content generated by the users, The US Supreme Court is debating about the responsibility of social media about what their users publish in their profiles (Paul, 2024). The outcome could have a deep impact on the liability of AI developers, as the case of AI generative, when users employ the application to create fake images of real people, using this material for political propaganda or sexual exploitation. The European Union has provided big steps regarding the regulation of AI. In March 2024, the EU

Parliament approved the AI act whose main objective is to provide protections to the development and use of AI. This act limits the use of AI for biometric identification, particularly in the case of law enforcement, prohibits the use of AI to commit frauds or manipulations against users, and gives the right to users to present complaints about potential abuses from corporations and receive further information about the products. At the same time, the EU is about to approve the AI Liability Directive, which will complement the AI act and provide a set of rules regarding the responsibility of AI developers of the potential damages that their apps could cause their users. Regarding this issue, some authors have argued that besides the regulations "the problems in proving fault and causality when errors in AI systems are difficult to foresee for producers, and monitoring duties of users are difficult to define" (Buiten et al., 2023, p. 1). But what is imperative about the ethical and legal affairs of AI, is to consider, in actual and future regulations and frameworks, some key principles such as human dignity and safety above economic profits (Thompson, 2021).

Aims and Scope

This volume discusses the social impact of current advances in artificial intelligence in Latin America, and the potential ethical and legal controversies caused by this technological adoption process. While in recent years there have been plenty of publications about this topic in the Northern Hemisphere, the case of USA and European academia, further research that focuses on other regional realities, such as in Latin America is needed. The galore of chapters presented in this book explore these controversies from the Latin American perspective, they contribute to the global debate about the incorporation of AI into people's daily lives but considering the cultural and social particularities of this region. A territory that besides the technological lacks and the economic and social issues, is eager to incorporate these disruptive technologies to face traditional problems such as criminality, accountability and the lack of educational coverage and inclusion.

This edited collection analyzes, from a critical and comparative approach, the potential benefits of using artificial intelligence to surpass traditional social and economic problems in Latin America, but to understand, at the same time, the perils and barriers derived from the adoption of this technology. Such as the lack of proper legal frameworks and the latent ethical conflicts of using these applications, particularly considering the protection of users from the mistreatment of private data, or the use of deep fake to promote misinformation. In addition, there are challenges to introducing this technology in a region such as Latin America with deep economic and technological disparities, not just locally, among rural and urban communities, but at a regional level, between more technological developed countries such as the case of Brazil and Mexico, and those Latin American countries that are now suffering from economic depression and/or political instability as the case of Venezuela or Cuba. But we cannot ignore the traditional economic and technological disparities between the North and the South. Where the North usually stands as the hub of innovation and richness and the South is the customary receptor of digital devices, applications and contents produced under the cultural logic that is not necessarily compliant with developing countries' interests and needs.

At a regional level, this growing interest for the AI development at a private and public sphere has appeared in a landscape shaped by political polarization among the left and right and a digitally interconnected society, where social media and mobile devices have become the main sphere to debate about public issues and ludic and reactive activities. Regarding the economy, there has been a huge rise of inflation particularly after the pandemic. The case of Argentina and Venezuela are remarkable because of the hyperinflation their citizens must deal with. There are great challenges regarding the management of internal migration flows of Cubans, Haitians, Venezuelans, Centro Americans, and refugees from distant countries such as India and Russia, that must travel across the Darien Gap into the US-Mexican Border. Countries that have low rates of criminality such as Ecuador and Argentina are now facing rising levels of insecurity partly because of the consolidation of international criminal gangs that operate across the region (Barreiro Santana, 2024).

These problems and challenges, however, have not limited the interest in AI in the region. On the contrary, in recent years, the expectations for Artificial Intelligence has grown. There is great expectation for applying the AI potential to face endemic problems in the region: poverty, crime, water supply, traffic, and pollution in big metropolises such as Mexico City, Buenos Aires, or São Paulo (Pombo et al., 2020). Governments in Brazil, Argentina and Mexico are trying to maximize the advent of AI by crafting new policies and installing novel laboratories (MIT Technology Review Insights, 2020; OECD & CAF – Development Bank of Latin America and the Caribbean, 2022). But far beyond these promising actions, the region faces strong limitations: the traditional problems associated with the digital divide that restricts the universal access to information technologies and therefore relents future developments on the IT sector, the lacking of legal and ethical frameworks, not just to incentivize the development of AI, but to set regulations that protects users' privacy when using AI apps, watch over potential illegal uses of AI to commit crimes such as frauds and to instigate social polarization, encourage innovation and investments (Baptista, 2023) and protect intellectual property rights (Díaz Charquero & Rangel, 2023). Probably one of the most difficult issues regarding AI regulation in America Latina, is promoting homogenization of legal frameworks (Porcelli, 2020; Durán, 2024). This not only has to be with upgrading actual norms in diverse legal fields such as criminal law, intellectual property, privacy and civil liability, and labor rights; but to operationalize regional cooperation to uniform technological standard and public policies to impulse AI's development in a sustainable and responsible way (Weck & Salazar Solis, 2020; Contreras Vásquez, 2024).

But it is also important to set regulations that not only incorporate ethical and legal principles shared by international organizations but to respond to Latin America's regional realities. In 2018, the European Commission established that "AI systems must not interfere with democratic processes or undermine the plurality of values and life choices central to a democratic society" (European Commission, 2018, p. 14). A further analysis on this extract raises some questions as there is no clarity about what should be understood by "democratic values"? and, what is more important, are we sharing the same values and principles? Regarding this issue, OECD in 2019 delivers a series of ethical principles to promote the development of trustworthy AI; among them, it is possible to highlight (a) sustainable development, (b) human rights and democratic values, (c) accountability and (d) transparency. Aligned with those principles, the UNESCO (2021) established a series of ethical guidelines to encourage research about AI but in a lawful and proactive way. Here are some key issues: the respect and protection of human rights, promoting equity and dignity, sustainability, responsibility and accountability. While it is important to highlight these ethical frames, it is also imperative to embrace them into Latin American legal frameworks from a critical and horizontal scope.

Among the most relevant topics in academic literature about AI, privacy stands as the most important. This is not only because artificial intelligence nourishes data, open and private, to learn and create content but also due to previous controversies such as the Cambridge Analytica electoral scandal in 2016, which was based on employment of users' data to manipulate political preferences. Actually, there is a galore of smart gadgets that employ AI-powered virtual assistants, the case of Microsoft's Cortana or Apple's SIRI, or smart toys, which have been under the spotlight due to the potential invasion of user's privacy as they record private conversations (Albuquerque et al., 2020). Privacy in the Age of Artificial Intelligence has become a major issue, for economic reasons but also due to the potential misuse of private data (Aguerre, 2020). This is particularly sensitive regarding the young population.

Minors are especially vulnerable to the exposition of their private life on social media and smart gadgets. Recent cases on the matter well illustrate this exposure. In Mexico and Argentina there have been incidents where personal images have been manipulated using Generative AI, in those cases the offenders use AI software to create false pornographic content using their female classmates' photos, all of them underage (Guardia Nocturna, 2023; Parisi, 2024). Most of the countries in the region have regulations that protect minors' privacy. In Mexico, for example the Ley General de los derechos de niñas (General Law for the Rights of Children and Adolescents), in Argentina, the Ley de Protección Integral de los Derechos del Niño (Law on Protection of the Rights of the Child) and Brazil with the Estatuto da Criança e do Adolescente (Child and Adolescent Statute). It is mandatory to update those regulations particularly considering the liability of AI corporations from privacy abuses but also from potential misuses from private users, as the cases mentioned before. Public efforts should be invested in creating legal and technological safeguards to protect users in general, but minors in particular when interacting with Generative AI. This protection should be applied at least in two ways. To avoid damages to users but also to limit the negative effects to third-parties. Other privacy issues will require further protections also, such as the use of facial recognition cameras to manage migration flows; the implementation of those devices requires the set of procedural specific guidelines to safeguard the data from misuses that could cause racial bias or false incrimination (Venturini & Garay, 2021). Or the implementation of Health chatbots to assist patients at hospitals or at home. It is not just to secure the accuracy of the information provided by the AI to the users, but to guarantee the protection of sensible users' data such as the results of diagnostic tests or the medical treatments (Medinaceli Díaz & Díaz Silva Choque, 2021).

Along with the protection of privacy, other legal topics, including intellectual property rights and the use of artificial intelligence in the creative process, requires further attention. Some artists and media corporations are starting to ask for more regulation on the use of AI and its potential impact on the media sector. Not only creating stuff but also replacing actors and writers. In Latin America, as most of the countries have confirmed the WIPO' agreements on protecting intellectual property rights (IP), their regulations need to be updated to set a new legal framework regarding the works created by AI software. And, if this is the case, consider liability or economic compensations when protected material is employed during the creative process (Magnani, 2024).

Organization of the Book

This volume delivers insightful analyses on cases related with the ethical and legal dilemmas of incorporating artificial intelligence in diverse socio-economic fields in Latin America. Topics associated with the employment of the application such as ChatGPT in the production of news, the social imaginaries associated with AI, the incorporation of AI in the labor market, the use of biometric and AI systems to deliver migrants aid relief, and the impact of using AI for educational purposes, among others topic. But these works focused, at the same time, on actual and future limitations, controversies and problems caused by this technology: the potential threats to the regional economy and prospective loss of jobs, the invasion of privacy, social exclusion, and the ethical dilemmas linked to the incorporation of AI to protect national borders.

This volume is divided into two sections. The first one, **"Considering AI in the Private Sphere,"** debates about social uses of artificial intelligence from the perspective of the citizens. In Chap. 2, "Artificial Intelligence, Intellectual Property, and the Labor Precarization of Journalists in Latin America," David Ramírez Plascencia studies how applications based on AI generative will introduce changes in the news-production process, how information is created and published online, and in what way AI will potentially bring challenges for the journalists' labor conditions. As Ramírez Plascencia argues, the final outcomes will show recent achievements about how AI generative present important technical challenges not only for journalists but for newspapers as well, particularly regarding the protection of their intellectual property rights. Chapter 3 "From Epistemological Foundations over Futuristic Speculations to

Fact-Based Concerns: Evolving Discussions on Artificial Intelligence in Argentina, Colombia, and Mexico (2014–2024)," is dedicated to understanding how media coverage about AI in Latin America conditions people's perspective on these technologies. Throughout the analysis of news and editorial opinions, Fátima Ávila-Acosta, Lucía Morales Lizárraga, and Jan Nehring identify prevalent topics, and analyze the hopes and concerns embedded in the media narratives. Narratives that, in some way, influence the peoples and government's attitudes toward adopting AI applications into their daily activities and processes. Chapter 4 "The Algorithmic Tyranny in the Gig Economy: National Strategies and Policy Implications in the Latin American Context," Alisa Petroff ponders the transformative potential of AI in Latin America but also the infrastructure and investment barriers that limits the process of adoption in the region. She centers on the economic sector, analyzes national strategies, their strengths and weaknesses. She studies the impact of AI in the Latin America labor market, how this technology affected the creation of new employment but also the transformation and elimination of other labor places, and how AI is transforming the marketplace of online freelancers, workers who are contracted across online platforms that operate under recommendation systems. This section closes with Chap. 5. "Mapping the Future of AI Regulation in Latin America: Civil Society Perspectives on Brazil's Pilot AI Regulatory Sandbox," a deep analysis on the role of civic engagement in the consolidation of AI regulation. In this work, Kenzo Soares, provides evidence about the actions that civil organizations are using to widen their voice in the legislative debate to create legal frameworks around artificial intelligence in Brazil, a country that in recent years presents important investments in the AI sector such as the establishment of super computers dedicated to the development and use of AI. This work focuses on the figure of regulation sandboxes, which are controlled settings arranged by regulators that allow testing novel AI developments under temporary relaxations of certain normativity, measuring the potential risks of the systems.

The second section, **Challenges and Promises of AI in the Public Sector,** centers on the actions undertaken by some Latin American governments to employ AI developments to solve different public issues from public administration, homeland security, and education, but at the same time, how they ponder the potential legal gaps and ethical dilemmas of introducing AI to deliver public services, improve government internal functioning and diminish social exclusion. Chapter 6, "Ethical and Legal

Dilemmas of National AI Policies in Latin American Countries," by Rosa María Alonzo González, analyzes the diverse ethical and legal approaches undertaken by Latin American countries to foster the development and incorporation of AI into the public administration, but also to encourage economic growth and social inclusion. In her research, Alonzo González dissects the main national AI policies created in the region, their ethical principles, goals, potentialities, and legal gaps. She also criticizes their limits and proposes some actions to increase the impact of this promising technology in the region. In Chap. 7, "The Use of Biometrics and AI for Border Control and Its Impact on Human Rights," Jezabel Pérez and Natalia Brzezinski analyze the case of smart borders, where the use of AI, biometrics and drones is replacing human surveillance. However, this technologization of border control conveys a series of legal gaps and the violence of human rights. This problem is particularly relevant considering the capture of biometric data by shelter countries such as Brazil and the employment of AI mobile applications such as CBP One by the US government to manage migration crises in Latin America. This chapter explores how besides security and humanitarian justification the operation of these technologies stands a potential violation of migrants' human rights who are forced to provide their personal information and biometric data without legal or technological safeguards.

Chapter 9, "Ethical and Educational Dilemmas of AI in Latin American Higher Education Institutions: Persistent Challenges and Inquiries" by Jairo Alberto Galindo-Cuesta, Rubén Yáñez Reyna, Paola Mercado Lozano, focuses on the main strategies developed by educational institutions in the region to incorporate the diverse AI applications, particularly those based on Generative AI, as a strategic action to improve education performance. This work surveys the main features and contexts of implementation of ethical guidelines in Latin American universities about the use of AI for teaching, learning and promoting educational innovation. The aim is to describe the main transversal elements of these policies and clarify the main barriers and challenges to incorporate Generative AI in didactic and pedagogic strategies. Finally, the volume closes with a reflection on the key contributions of the individual chapters. Their arguments are about the benefits of AI but also the ethical and legal risks in the governmental, economic and education sectors. Chapter 9 "The Quest for a Responsible Use of AI in Latin America", by Alonzo González and Ramírez Plascencia, also delivers a critical approach to the incorporation of AI in Latin Americans' daily lives, and how to create the proper settings to promote a responsible use of artificial intelligence at a national and regional level. According to the authors, fomenting a responsible use of AI is a collective task, therefore requiring the collaboration of all actors: citizens, authorities, corporations and non-governmental organizations.

Final Remarks

This volume presents a galore of studies that approach the vibrant but complex process of artificial intelligence adoption in Latin America. As cases analyzed in this book have shown, some of the current AI developments and policies entail not only IT infrastructure challenges but also the absence of legal and ethical frameworks at a local and regional level. But what is imperative is the involvement of civic organizations in the composition of norms that potentially affect users' rights and discourage the incorporation of AI among groups that have been traditionally excluded. More horizontal and critical activism from developers, practitioners and communities are needed to embrace this new wave of technological shifts in a more equitable and inclusive way.

While some applications such as ChatGPT, Google Gemini, and Microsoft Copilot are already accessible to the wider public, we are still pondering their real influence and impact not only in the ludic or educational aspects but also on the legal scope. There are still pending legal cases to be heard in court as the manipulation of real images using AI generative to commit frauds or sex crimes. The introduction of AI in the media sector is the source of diverse apprehensions as writers, translators or journalists consider these applications not as tools that could help them to improve their labor but unfair competitors that will eventually replace them. However, this sector is not the only menaced by the incorporation of AI, as many labor activities, such as customer service, have been digitized, especially during the pandemic and they are also susceptible of being replaced by AI apps briefly. While other issues such as the use of AI to surveil and combat crime or in military conflicts looks futuristic, many of them, due to their cheapening and improvement, as the case of drones or face recognition systems, are becoming more common among the public forces and in the control of national borders across the continent. While this volume does not attempt to be exhaustive, its intention is to promote the public debate about AI's promises and barriers in the light of cultural disparities but also potentialities in a region where 30% of the population does not reach 25 years of age. It aims to provide critical studies on key economic and social sectors in a region where the AI is not just an imaginary conjecture but also an incipient reality.

References

Agrawal, A., Gans, J., & Goldfarb, A. (2018). *Prediction Machines: The Simple Economics of Artificial Intelligence.*

Aguerre, C. (Ed.). (2020). *Inteligencia Artificial en América Latina y el Caribe. Ética, gobernanza y políticas.* Universidad de San Andrés.

Albuquerque, O. de P., Fantinato, M., Kelner, J., & de Albuquerque, A. P. (2020). Privacy in smart toys: Risks and proposed solutions. *Electronic Commerce Research and Applications*, 39, 100922. https://doi.org/10.1016/j.elerap.2019.100922

Baptista, D. (2023, July 10). ChatGPT drafts law as Latin America seeks to regulate AI. *Context.* Retrieved from https://www.context.news/ai/chatgpt-drafts-law-as-latin-america-seeks-to-regulate-ai

Barreiro Santana, K. (2024). Crimen organizado, conflictos fronterizos, lavado de activos y ciberespacio: Desafíos actuales en la región. *Estado & comunes, 1*(18), 203–208. https://doi.org/10.37228/estado_comunes.v1.n18.2024.343

Brambilla, I., César, A., Falcone, G., & Gasparini, L. (2023). The Impact of Robots in Latin America: Evidence from Local Labor Markets. *World Development, 170*, 106271. https://doi.org/10.1016/j.worlddev.2023.106271

Brennen, S., Howard, P. N., & Kleis Nielsen, R. (2018). *UK Media Coverage of Artificial Intelligence Dominated by Industry, and Industry Sources.* The Reuters Institute for the Study of Journalism.

Brynjolfsson, E., Li, D., & Raymond, L. R. (2023). *Generative AI at Work* (Working Paper 31161; Working Paper Series). National Bureau of Economic Research. https://doi.org/10.3386/w31161

Buiten, M., de Streel, A., & Peitz, M. (2023). The Law and Economics of AI Liability. *Computer Law & Security Review, 48*, 105794. https://doi.org/10.1016/j.clsr.2023.105794

Chen, B. X. (2023, June 30). Don't Use A.I. to Cheat in School. It's Better for Studying. *The New York Times.* Retrieved from https://www.nytimes.com/2023/06/30/technology/ai-chatbot-study-aid.html

Contreras Vásquez, P. (2024). Convergencia internacional y caminos propios: Regulación de la inteligencia artificial en América Latina. *Actualidad jurídica iberoamericana, 21*, 468–493.

Dauvergne, P. (2021). The Globalization of Artificial Intelligence: Consequences for the Politics of Environmentalism. *Globalizations, 18*(2), 285–299. https://doi.org/10.1080/14747731.2020.1785670

Díaz Charquero, P., & Rangel, V. (2023). *Informe sobre Políticas de Inteligencia Artificial y Derechos de Autor en América Latina*. Fundación Karisma. Retrieved from https://web.karisma.org.co/informe-sobre-politicas-de-inteligencia-artificial-y-derechos-de-autor-en-america-latina/

DiMatteo, L. A., Poncibò, C., & Cannarsa, M. (Eds.). (2022). *The Cambridge Handbook of Artificial Intelligence: Global Perspectives on Law and Ethics* (1st ed.). Cambridge University Press. https://doi.org/10.1017/9781009072168

du Boulay, B., Mitrovic, A., & Yacef, K. (Eds.). (2023). *Handbook of Artificial Intelligence in Education*. Edward Elgar Publishing Limited.

Durán, R. (2024). Índice Latinoamericano de Inteligencia Artificial. CENIA/ILIA. https://blog.okfn.org/es/2023/08/14/indice-latinoamericano-de-inteligencia-artificial/

Elish, M. C., & Hwang, T. (2015). *Praise the Machine! Punish the Human! The Contradictory History of Accountability in Automated Aviation*. Data &Society Research Institute. Retrieved from https://www.datasociety.net/pubs/ia/Elish-Hwang_AccountabilityAutomatedAviation.pdf

European Commission. (2018). *High-Level Expert Group on Artificial Intelligence Draft Ethics Guidelines for Trustworthy AI*. European Commission.

Fast, E., & Horvitz, E. (2017). Long-Term Trends in the Public Perception of Artificial Intelligence. *Proceedings of the AAAI Conference on Artificial Intelligence, 31*(1), 1. https://doi.org/10.1609/aaai.v31i1.10635

Floridi, L. (Ed.). (2021). *Ethics, Governance, and Policies in Artificial Intelligence* (Vol. 144). Springer International Publishing. https://doi.org/10.1007/978-3-030-81907-1

Guardia Nocturna. (2023, October 13). Había sido puesto en libertad por "falta" de pruebas, a pesar de que en su ipad se habían localizado alrededor de 20 mil fotografías #guardianocturna. *Instagram*. https://www.instagram.com/p/CyWgU3qum28/

Haney, B. S. (2020). Applied Artificial Intelligence in Modern Warfare and National Security Policy. *Hastings Science and Technology Law Journal, 11*, 61.

Hundt, A., Agnew, W., Zeng, V., Kacianka, S., & Gombolay, M. (2022). Robots Enact Malignant Stereotypes. In *2022 ACM Conference on Fairness, Accountability, and Transparency* (pp. 743–756). https://doi.org/10.1145/3531146.3533138

Katz, Y. (2020). Artificial Whiteness: Politics and Ideology in Artificial Intelligence. In *Artificial Whiteness*. Columbia University Press. https://doi.org/10.7312/katz19490

Liao, S. M. (Ed.). (2020). *Ethics of Artificial Intelligence*. Oxford University Publication.

Magnani, L. B. (2024, February 13). Direitos autorais e inteligência artificial. *JOTA Jornalismo*. https://www.jota.info/artigos/direitos-autorais-e-inteligencia-artificial

Medinaceli Díaz, K. I., & Díaz Silva Choque, M. M. (2021). Impacto y regulación de la Inteligencia Artificial en el ámbito sanitario. *Revista IUS, 15*(48), 77–113.

MIT Technology Review Insights. (2020). *The global AI agenda: Latin America* (The global AI agenda series). MIT.

Murphy Kelly, S. (2024, January 26). Las imágenes explícitas de Taylor Swift ponen el foco en el lado más oscuro de la inteligencia artificial. *CNN*. Retrieved from https://es-us.vida-estilo.yahoo.com/difunden-im%C3%A1genes-expl%C3%ADcitas-taylor-swift-010650757.html

OECD & CAF Development Bank of Latin America. (2022). *The Strategic and Responsible Use of Artificial Intelligence in the Public Sector of Latin America and the Caribbean*. OECD. https://doi.org/10.1787/1f334543-en

Panditharatne, M., & Giansiracusa, N. (2023). *How AI Puts Elections at Risk—And the Needed Safeguards*. Brennan Center for Justice. Retrieved July 22, 2023, from https://www.brennancenter.org/our-work/analysis-opinion/how-ai-puts-elections-risk-and-needed-safeguards

Parisi, M. (2024, October14). Escándalo en un colegio de San Martín: Denuncian a un alumno que vendía fotos manipuladas con IA de sus compañeras desnudas. *Clarín*. https://www.clarin.com/policiales/escandalo-colegiosan-martin-denuncian-alumno-vendia-fotos-manipuladas-ia-companeras-desnudas_0_hqEbfd1Plm.html

Paul, K. (2024, February 26). The US supreme court appears skeptical of social media content moderation laws. *The Guardian*. Retrieved from https://www.theguardian.com/technology/2024/feb/26/social-media-moderation-us-supreme-court-case

Pombo, C., Cabrol, M., Alarcón, N. G., & Ávalos, R. S. (2020). *fAIr LAC: Responsible and Widespread Adoption of Artificial Intelligence in Latin America and the Caribbean*. IDB. Retrieved from https://publications.iadb.org/en/fair-lac-responsible-and-widespread-adoption-of-artificial-intelligence-in-latin-america-and-the-caribbean

Porcelli, A. M. (2020). La inteligencia artificial y la robótica: Sus dilemas sociales, éticos y jurídicos. *Derecho global. Estudios sobre derecho y justicia, 6*(16), 49–105. https://doi.org/10.32870/dgedj.v6i16.286

Ramírez Plascencia, D. (2020). Tropicalizing Frankenstein in Latin America. A Tale about Promising Technologies and Apocalyptic Robots. *Athenea Digital. Revista de Pensamiento e Investigación Social, 20*(2), 2. https://doi.org/10.5565/rev/athenea.2647

REPEF. (2023). *Aportes feministas en Inteligencia artificial: Justicia epistemológica y ética de la igualdad* (Política Exterior Feminista). Ministerio de Asuntos Exteriores de Argentina.

RT International. (2017, September 1). 'Whoever leads in AI will rule the world': Putin to Russian children on Knowledge Day. *RT International*. Retrieved from https://www.rt.com/news/401731-ai-rule-world-putin/

Sharkey, A., & Sharkey, N. (2021). Sunlight Glinting on Clouds: Deception and Autonomous Weapons Systems. In A. Henschke, A. Reed, S. Robbins, & S. Miller (Eds.), *Counter-Terrorism, Ethics and Technology: Emerging Challenges at the Frontiers of Counter-Terrorism* (pp. 35–47). Springer International Publishing. https://doi.org/10.1007/978-3-030-90221-6

Sheikh, H., Prins, C., & Schrijvers, E. (2023). *Mission AI: The New System Technology*. Springer International Publishing. https://doi.org/10.1007/978-3-031-21448-6

The Economist. (2024a, February). How cheap drones are transforming warfare in Ukraine. *The Economist*. Retrieved from https://www.economist.com/interactive/science-and-technology/2024/02/05/cheap-racing-drones-offer-precision-warfare-at-scale/

The Economist. (2024b, February). The end of the social network As Facebook turns 20, social apps are being transformed. *The Economist*. Retrieved from https://www.economist.com/leaders/2024/02/01/the-end-of-the-social-network

Thompson, S. J. (Ed.). (2021). *Machine Law, Ethics, and Morality in the Age of Artificial Intelligence*. IGI Global. https://doi.org/10.4018/978-1-7998-4894-3

Venturini, J., & Garay, V. (2021). *Facial recognition in Latin America*. AlSur / NED.

Weck, W., & Salazar Solís, L. A. (2020). *Inteligencia artificial en Latinoamérica*. Konrad Adenauer.

PART I

Considering AI in the Private Sphere

Artificial Intelligence, Copyright, and Media: An Analysis of Journalism in Latin America from an International Perspective

David Ramírez Plascencia

Introduction

As the impact of Artificial Intelligence spreads into the economic, health, education, and military sectors, the media industry looks expectant amid an optimistic aura over AI's potential to automate the process of content creation and delivery. But there are also suspicions toward latent adverse impacts: copyright infringements from AI developers and potential loss of jobs: activities that have been traditionally made by humans could be eventually replaced by AI, such as translators, scriptwriters, or digital content creators. Recently, some newspapers complained about the use of AI

D. Ramírez Plascencia (✉)
Centro Universitario de Guadalajara (CUGDL),
Universidad de Guadalajara (UdeG),
Guadalajara, Jalisco, Mexico
e-mail: davidram@udgvirtual.udg.mx

D. Ramírez Plascencia, R. M. Alonzo González (eds.), *Ethical and Legal Dilemmas of Artificial Intelligence in Latin America*,
https://doi.org/10.1007/978-3-031-86540-4_2

in the entertainment industry such as to compose music (Europa Press, 2023) or write literary works (Israde, 2024). In both cases, the industry claims for more controls and regulations toward the employment of artificial intelligence to elaborate creative content. Besides the disruptive nature of AI, these mixed feelings about incorporating novel technologies into the mass media are not new.

Over time, the media industry has been reluctant to introduce new technology to their distribution processes. Such as the case of the music industry with Napster and iTunes Music in the late 1990s and early 2000s or the cinema corporations with the appearance of Netflix and the proliferation of the VOD services. Since the popularization of the Internet in the mid-1990s of the past century, traditional media such as radio, music, the publishing sector, TV, and cinema have experienced a long-lasting transition from the analogic structure into the digital format. This transformation has brought a galore of benefits: capabilities to reach a wider public, reduction of production costs, as they no longer depend on the physical format, and the diversification of the revenue sources. But this technological shift also implies challenges, particularly regarding the system through which companies and artists merchandize, distribute and sell their content. This has been also the case of journalism, which has experienced in the past decades a long transition from the print-paper into the digital layout, a shift shaped by a strong presence in social media, the rising of mobile consumption of news, automatization, and now the arrival of Generative AI to create content. The advantages brought by digital media in the news production process are undeniable: the diversification of information sources, flexibilization of management procedures, enhancement of the communication channels to spread and access the news, interactivity among the public and the media, and the search for alternative sources of financing: subscriptions, patronages and digital ads, among others. The process of digitization of news, though, has caused other adverse effects: journalists' labor precarization, legal issues related to copyright, and a rising competition with social media platforms and digital-content creators.

In Latin America, besides the growing importance of the free press to protect democratic institutions, denounce community issues, and give voice to the underrepresented and marginalized groups, the working conditions of journalists are among the worst. Many of them work as freelancers with few or no rights; there are even cases in which they make collaborations without payment in exchange for visibility. These circumstances contrast with the competitiveness and diverse skills to which journalists must be familiar with to comply with the labor market demand. In addition, working as a journalist is considered as a profession of high risk in Latin America where correspondents are murdered or coated to remain silent or compelled to support the government and not investigate criminal organizations' felonies. This chapter will focus on understanding the potential benefits but also the latent risks brought by the introduction of artificial intelligence for journalism in Latin America. This chapter will analyze issues related with the iteration and the use of AI generative to produce news and digital content. While AI could stand as a great support to journalists, this incorporation also could represent the replacement of human intervention in the news production and the consequence of the loss of jobs, whose activity depends largely on intellectual work. As this chapter argues, most of the debates are not only because of the introduction of technology but also because of the lack of efficient legal frameworks to protect authorship and provide a more equal retribution for the media and the journalists' labor, and the decline of public appreciation of journalism as a public good and a democratic pillar.

Digital Journalism: From Paper to Bits

While it is possible to date the beginning of digital journalism up to the early 1980s, it was during the mid-1990s, with the popularization of the Internet, that big news corporations such as Le Monde, The Telegraph or The New York Times started publishing their news online (McCormick, 2023). The first decades of the new millennium witnessed a long but certain decline of the print press and the consolidation of the digital journalism (Charon, 2010). People shifted from buying newspapers to the consumption of news online (Newman, 2023), by now, social media has consolidated as the main source of news information. Computers, Internet and, later the arrival of social media, enhanced the possibilities of journalists that were able to access massive sources of information and whose work could potentially reach a wider public across the globe. Technology has also enabled the launch of digital platforms enriched with multimedia elements that allows interaction with the public, flexibility and the production of content in real time. This technological shift stood a big challenge to journalists also who were compelled to mastering novel software and devices, but also to modify their traditional forms of doing their job: from a rigid and vertical model up to a more flexible and scalable method, where traditional media corporations have lost the monopoly of information and novel actors such as Alphabet (Google) and Meta (Facebook) which have consolidated as the new media moguls (Iordanidou et al., 2022). The innovation brought by the process of the digitization of the news media sector has caused some contrasting impacts as well: (a) the closing of newspapers (print and digital) since they were able to continue dwelling with the fixed costs of producing news and (b) The long-lasting precarization of journalists' labor conditions not only in Latin America but also overseas.

The End of the Print Press Model

Around 2010, the traditional newspaper business looked daunting. There were at least three indicators that highlighted this crisis (Kaiser, 2014). (1) The revenues dropped and the spending on advertising greatly move to the Internet, (2) the numbers of professionals dedicated to journalism has sustainably declined in the USA (3) The sources of information diversified, as there are other companies and individuals, besides newspapers, that are now competing in the news business. These factors also manifested in Europe as the print-press model entered in a constant crisis, struggling to survive among a changing context (Langley, 2012; Watson, 2024).

As the process of digitization advanced, newspapers across the world were compelled to apply severe spending cuts or close as they could not meet their fixed costs of producing news. Besides difficulties to remain in the print-press business, many newspapers continued releasing their paper edition, but it was the 2020 pandemic that gave the "coup de grace" to many of those efforts (Carvajal, 2020). However, as some data reveals, digitalization was not the last resort for newspapers. Just before the arrival of the 2020 pandemic, 360 newspapers were out of business in the USA, this number encompasses the total of 2500 that have closed since 2005 (Simonetti, 2022). According to the latest report of The State of Local News (Muse Abernathy, 2023), this negative tendency remains as a media of 2.5 newspapers closed per week in the United States in 2023. This situation, however, not only relies on the lack of investments, as The Washington Post and The Los Angeles Times have been acquired by billionaires who have put in some resources. Yet these media have not shown to be profitable and still suffering financial losses (Mullin & Robertson, 2024).

These tendencies are also present in Latin America: a sustained decline of printed newspaper sales and the rise of digital news (AFP, 2020).

However, amid this technological transition, there are other endemic problems that have also affected the industry in the region (Salaverría et al., 2019). One of them is censorship and repression from authorities. Several newspapers in Venezuela were forced to close mainly because repression from Hugo Chávez and his successor Nicolás Maduro (Editorial La República, 2018). Some medias in Bolivia such as "Página Siete" and "El Periódico" in Guatemala closed due to the same issue (FRANCE 24, 2023). Another organic problem is violence, as journalists suffer from verbal menaces up to physical violence or even murder. In countries such as Mexico or Colombia, being a journalist is becoming a high-risk profession; the drug cartels and guerrillas usually threaten or kill journalists to stop researching corruption or crime cases. According to the organization Artículo19 (2022), in between 2000 and 2022, there have been 164 professionals assassinated. As it has happened in other latitudes such as The United States or Europe, the pandemic contributed to accelerate the migration to the digital format in Latin America, as it was difficult to continue distributing the paper version. This radical and quick shift caused a decrease of revenues forcing newspapers to find alternative forms of commercialization, such as having more presence on the Internet, particularly on social media (Mioli, 2021). This context negatively impacted not just the newspapers but a whole production chain from local book and newspaper stores to street sellers, but also paper producers, printers, etc. (AFP, 2020).

Journalists' Labor Precarization

Journalists overseas have suffered a sustained process of precarization. As the local and international media corporations started applying spending cuts, the first "victims" of this economic shift were the employees. In Spain for example, some newspapers, as the case of El Pais, recurred to the use of regularizing employment measures (Expediente de Regulación de Empleo -ERE-). In 2012, El País fired 129 employees, many of them with more than 20 years of experience in the company (Redacción AN, 2012). Other massive layoff cases illustrate well another problem related with media financing: the high dependency to public investments. The case of Télam, in Argentina, which loses about 40% of its workforce (354 employees) in 2018 due to, as declared by the Argentinian government, a lack of economic sustainability (Molina & Centenera, 2018). This dependence on public money caused labor instability, since the operations stand on the availability of public resources and the alternance of the ruling party, which usually uses this media for political and ideological purposes, restricting journalists' free expression, and imposing a determinate editorial line. The post-pandemic context did not bring better expectations for journalists' labor market, but massive layoffs and the closing of businesses remained a constant. As the case of the Washington Post that fired 240 employees in 2023 and Los Angeles Times, which at the beginning of 2024, dismissed about 20% of the workforce or 115 employees (Beauregard, 2024).

These mass layoffs translated into low salaries as well. In 2017, some testimonies illustrated well the impoverishment of labor conditions (Herrscher, 2017): 100 dollars or less per an article (In USA and Latin America), and salaries that barely surpass the 300 dollars per month for a journalist in Colombia. In Mexico in 2019, the average salary did not reach 400 dollars, at the price of nearly 20 Mexican pesos per one dollar (Rivera Escobedo, 2020). Adverse labor conditions in Mexico continue, journalists are compelled to fulfil extended journeys in exchange for low wages (Juárez, 2023). But this did not reflect the actual state of the labor market, since many of them work as freelancers, with no labor rights (Blanco-Herrero et al., 2020; Gollmitzer, 2024). The latest stats in Mexico showed that nearly 34% of the workforce was informal (Data México, 2024). Evidence from Brazil (FENAJ, 2023) and Peru (Sarmiento, 2024) show the same adverse labor panorama. There are even cases where journalists work for free in exchange for public visualization (Pennacchio, 2023).

Ironically, besides precarization, journalists have been forced to capacitate and acquire novel labor skills: from basic abilities to surf and search information on the Internet and use office suites up to be able to publish on social media, and in many cases, capture and edit their own content: audio, video, make presentations, photos, infographics and animations (Agudelo González et al., 2022). But the acquisition of novel competencies does not necessarily translate into better salaries. On the contrary, as the fixed costs associated with producing news continues, the media made massive layoffs and reduced labor benefits for their remaining staff. To keep the ship afloat, the media corporations have undertaken three main actions (a) some employers continue working for the newspaper but as freelancers, (b) the remaining workforce acquired more responsibilities with less payment, and (c) many activities were automated. Computers were employed to handle and analyze the huge amount of data which serves as raw material for journalists, but also make the news production model efficient and reduce the time to prepare and share the information, using algorithms to evaluate and improve the quality of inside the editorial offices (LeCompte, 2015). Automatization also allowed publishing across platforms in multilingual versions in a faster and cheap form (Wölker & Powell, 2021).

The establishment of automation in journalism coincided with other factors such as the rise and consolidation of social platforms such as Facebook or X, formerly Twitter, that compete with traditional newspapers, putting pressure on the amount of information that should be published every day. In other words, newspapers have been compelled to publish more and more frequently (Wilding, 2018). While the irruption of automation has not entirely replaced the function of journalists (Darcy & Passantino, 2024), it paved the way to the introduction of more sophisticated apps based on Artificial intelligence, particularly Generative AI which capabilities stands, for many critics, a serious menace for the role of humans in the creative sector, not just for newspapers, but also in arts and entertainment.

AI, Intellectual Property Rights, and Journalism in Latin America

The incorporation of AI in Latin American journalism occurs in a context where social media has consolidated as the main source of information and entertainment. Meta (a company that owns Facebook and Instagram), Alphabet (Google) and TikTok practically embraced 70% of the global online traffic in the region (Castells & Rattel, 2024). TikTok has become very popular among Latin Americans, where users prefer consuming news in a "short-form video," one or two-minute extension than reading the entire news (Fletcher et al., 2024). Another study confirms the predominance of TikTok in Latin America, remarkably strong among the Z generation (Born amid 1997 and 2012). In this group, the use of TikTok reaches nearly 62% of their total consumption (Castellanos, 2024). Though newspapers are still operating their online portals, much of their traffic comes from internet searchers (mainly Google) or across social media profiles. Nonetheless, newspapers are starting to look for other spaces to monetize their content such as the production of podcasts, whose market consumption after the pandemic is in full growth in the region (Guttmann, 2024).

The incursion of Generative AI into journalism occurs in a context already shaped by automation (Salaverría & de-Lima-Santos, 2021). Daily activities such as capturing, indexing, categorizing, and presenting information, undertakings that were traditionally done by humans, are now carried out by computers. However, the composition of the news remains a journalist's responsibility who employs digital content such as videos, audios and text and virtual tools from word processors, audio and video editors to create and deliver the information. In Latin America, journalists have embraced the irruption of AI in journalism as an opportunity to acquire new abilities to search data and master applications such as ChatGPT to create enriched content, and media corporations foresaw AI as an opportunity to learn about consumer preferences and offer a more personalized content (CLIP/IMS, 2021). But the journalists' consideration about AI embraces mixed opinions. According to one recent report (LatAm Intersect PR, 2023), 82% of the respondents considered AI a helpful tool. However, about 30% of the surveyed in between 18 and 25 years old consider AI as a "threat." Another 32.6% mentioned that they have never used any AI app and just 35.6% use AI from every day to once a week. Whereas the potentialities of AI in journalism are in constant development, their usage in Latin America concentrates on proofreading documents, transcribing audios, translating news into diverse languages, and analyzing consumer trends, not just to deliver news according to preferences, but also digital ads among others (Zuazo, 2023). However, there are also other functions that could cause controversy such as the trustworthiness of the content created by AI or the potential replacement of human conductors by AI presenters (González, 2023).

In recent years, protests from the entertainment industry and cultural sector have become more frequent, they complain about the use of AI to elaborate creative works such as songs to literary works or even pieces of visual arts. These objections focus on two main issues: (a) the use of AI to replace humans, not just in routine jobs such as labeling and categorizing content, but in the creative process. This is the case of hundreds of radio DJs that were laid off by iHeartMedia in 2020 and substituted by AI software that mixes songs and schedules music. In 2024, Prime Video was publicly accused of using AI to replace voice actors in their Korean series, (b) the potential violation of copyright when AI uses protected material available online. This could be "learning" from copyrighted content to improve software capabilities or to create new stuff. The several cases covered by the media provide evidence of the growing importance of this issue among the creative industry (music, visual arts, the publishing sector, etc.). In 2023, Universal Music Group (UMG) asked streaming services such as Spotify and Apple to avoid allowing AI developing companies to access their musical catalog to train AI without the proper permission of the recording companies or the artists. In October 2023, the Hollywood writers and the Studios accorded a new agreement that set limits to the use of Generative AI to create series and movie scripts. It will be considered as a tool to improve the quality rather than a substitute for professional writers. However, other creative industries such as book publishers are not so enthusiastic with how AI uses their literary works to "learn." They argue that it is not about "fair use" but rather about stealing (Barber, 2023). In July 2023, The US Authors Guild sent a letter to AI corporations such as OpenAI and Microsoft asking them to obtain legal consent, give credits to the authors, and provide a monetary compensation to the writers for using their works to train their AI systems (Bedingfield, 2023).

Regarding the news media corporations, they have received AI, as it was possible to observe early in this chapter, as an opportunity to improve the news-production process to diminish fixed costs. However, this enthusiasm clashes with their reservations about how AI developers use their news to train their systems and deliver content (New Media Alliance, 2023). Practically, amid 2023 and 2024, the discussion about news, AI, and copyright has accelerated not only as a subject of debate in academia and economic forums but it has also materialized in lawsuits adducing intellectual property infringement. In 2023, Getty demanded the creators of Stable Diffusion, a generative AI model that forges realistic photos using content already published online, for violating copyright and trademark rights (Appel et al., 2023). In this context, diverse US newspapers are pondering creating an alliance to face the challenges of AI, particularly regarding the threats toward their intellectual property rights. Other media corporations, as the case of Associated Press, have already signed agreements with OpenAI regarding the use of their content (Golding, 2023). However, other newspapers have taken a different path. They are developing their own AI service to deliver news and other material to their readers (The Economist, 2024). It is, nonetheless, early to envisage the potential success of these actions.

Amid the actual discussion about AI and Intellectual property there are some key topics pending to be resolved, (a) one of them is the boundaries of "fair use," which is frequently adduced by AI developers to employ protected material, (b) to set new ways of licensing protected material to be used by AI creators. In this case, it is imperative that these procedures not only guarantee royalties for the media corporations but also to the journalists, since many times their content is used and republished in diverse platforms and formats. Though incipient, the discussion of these issues started to gain relevance in Latin American. There have been voices that claim for more control and regulations (Magnani, 2024; Israde, 2024; Paludi, 2024). Generally, these discussions recall global topics about the protection of copyrighted content, set limits to AI and regulate legal loopholes about authorship of AI creations.

However, these recent complaints and legal actions from the media industry about the use of AI should not be seen as unprecedented, but as continuation of early legal debates from the 2000s when the Internet started dominating the global media, and people began sharing files in Napster or Google created its news service (Google News). In 2005, the French news agency AFP (Agence France-Presse) sued Google because its service Google News published photos, headlines and other content without legal permission (El País, 2005). At that time, the discussion was not exclusively related with having permission to republish the content, arguing, "free use," since Google embedded ads in its news site. But that AFP wanted monetary compensation for this action, no matter that AFP members benefited from Google News, as many of the AFP portal visitors came by giving "click" to the headline in this platform. As it happened with other copyright infringement cases, it ended with a compensation agreement amid both sides in 2007. But this was not the end of the story, since only in 2021, 16 years after, Alphabet (Google) and AFP signed a commercialization deal that allows Google News to aggregate AFP's content (Dave, 2021). This case is an emblematic example of how difficult it could be to align legal and economic interests in the media industry. And how nowadays news corporations are struggling to diversify their financier sources beyond the advertising and subscription model. In Latin America, though emergent, the newspapers are starting to reinforce their presence online and finding alternative financing sources (Mioli, 2021), mainly from digital advertising (Navarro, 2024). This fact implies increasing the production of content and having a strong presence in social media and setting efficient actions to legally protect their content because of the ease that the Internet allows to share and modify the news and other multimedia elements. It is for these reasons that the arrival of AI may be a great ally to increase the content, the popularity and visibility of news but it also threatens the industry because of the lack of legal protection of newspapers' content and the quality and trustworthiness of the information produced using AI.

Concerning intellectual property regulation and the employment of AI in Latin America, it is possible to find some pending questions that, due to the lack of proper legal frameworks, are yet to be complied (Wachowicz, 2019). One of them is authorship. Is AI entitled to be considered the owner of the material it creates? (Lee et al., 2021) or should the ownership be adjudicated to the users that set the instructions to the software? Or to the software developers? Or even to those people or corporations that hold the copyright of the original works employed to generate the AI content. The problem about authorship, not only implies local efforts in every country, but internationally. As the WIPO (World Intellectual Property Organization) should be pronounced about this issue and promote updates if that is the case. Since by now, according to most national copyright regulations, including Latin American countries (Ávila Vallecillo, 2021), only humans could be entitled as authors of a protected work. Actually, there are diverse countries in Latin America that are working, or already have, on national policies to stimulate the development of Artificial Intelligence, such as Argentina or Brazil, but only some of them consider the intellectual property issue as a key issue (Díaz Charquero & Rangel, 2023). This absence is significant since as it was possible to observe in this chapter, copyright, along with private data integrity, is potentially one of the most important, but also controversial issues regarding the development of AI.

Conclusions

In 2019, Charlie Beckett published the report "New powers, new responsibilities: A global survey of journalism and artificial intelligence." The report presented some key asserts: (a) The robots are not going to replace journalists, (b) the machines will assume many daily works, (c) the AI will provide the journalists with new potentialities to undertake their job in an innovative way, and (d) AI will transform newsrooms into a "data-driven age" (Beckett, 2019, 01). Six years, as the advances in AI have impacted the sector, it looks like we are starting to observe some evidence of the real impact of artificial intelligence in journalism: IA news presenters, news created and delivered without human intervention and according to users' preferences in online portals, and even AI bots that delivers news and additional content. However, there are still debates about what can be done to join this new technological thread, how media and journalists should embrace the incorporation of AI? And how AI will affect media workers labor conditions? Some authors highlight the imperative of being prepared and adopting the AI in the newsroom, particularly in the case of local diaries (Marconi, 2020; Rinehart & Kung, 2022). What is certain about the introduction of disruptive technology in the newsroom, is that it has not been accompanied with effective legal strategies, particularly regarding the protection of intellectual property rights of media corporations and journalists.

The arrival of AI in the news media sector in Latin America occurs amid a context shaped by the dominance of big media corporations such as Alphabet and Meta, and the struggle of newspapers to find alternative sources of revenues. The long-lasting technological shift from print paper to digital format caused a sustained adverse labor situation to journalists who were, in a certain way, the first victims of the end of analogical financial model based on paper and the arrival of automatization and AI, as many of their functions have been replaced, or are in the transition of being substituted by artificial intelligence; or have suffered a substantial reduction of their labor conditions (Sanjurjo & de Frutos, 2022), in other cases, they have been compelled to work as freelancers or independent content-creators in social media.

This precarization, however, appears amid a process where the journalist has been successful acquiring novel competencies to harness first, the digital media model of producing news, but now also compelled to master AI software to manage information and produce more and better content. This paradox could be explained by three main factors, one of them economic (a) as newspapers depend more on digital advertising and visibility on social media. They, however, are not competing alone but with other information sources from journalist freelancers, influencers and big media conglomerates. This struggle to gain visitors and satisfy advertisers' demands, compelled newspapers to produce more and faster content at a lower cost, (b) another factor is technological, but it is strongly related to the previous one. In this quest to constantly generate content to feed their news portals and social profiles, newspapers have found in automatization and AI a strong ally that could undertake routine works such as cataloging and delivering information, and now even create the news and the photos with little or non-human intervention, (c) The last factor is ethical and relates to the consumers. As they do not consider newspapers as a relevant product that has the social function to report community issues or authority abuses, cases that big media corporations sometimes do not cover

(Kahn, 2024). People tend to forget the efforts and time invested in producing news and do not consider it as content that is worth to pay for (Álvarez Ugarte, 2022), but as another service that is given free of charge on the Internet, without considering other issues such as the accuracy of the information, the impartiality and the quality of the multimedia elements. But above all, citizens overlook that journalism is not like other media industries but a pillar of modern democracies and a public good (Sjøvaag, 2024).

References

AFP. (2020, August 7). La pandemia acelera el final de la prensa en papel, pero dispara la digital. *France 24*. Retrieved March 16, 2024, from https://www.france24.com/es/20200807-la-pandemia-acelera-el-final-de-la-prensa-en-papel-pero-dispara-la-digital

Agudelo González, L. E., Marta-Lazo, C., & Aguaded, I. (2022). Competencias digitales en el Currículo de Periodismo: Análisis de caso de una universidad Centroamericana. *Vivat Academia. Revista de Comunicación, 155*, 297–316. https://doi.org/10.15178/va.2022.155.e1393

Álvarez Ugarte, R. (2022). *Viabilidad mediática en América Latina: Hacia un periodismo sustentable* (Technical Report). Centro de Estudios en Libertad de Expresión y Acceso a la Información. Retrieved November 4, 2024, from https://dspace.palermo.edu:8443/xmlui/handle/10226/2389

Appel, G., Neelbauer, J., & Schweidel, D. A. (2023, April 7). Generative AI Has an Intellectual Property Problem. *Harvard Business Review*. Retrieved March 16, 2024, from https://hbr.org/2023/04/generative-ai-has-an-intellectual-property-problem

Artículo19. (2022, August 27). *Periodistas asesinadas/os—Artículo 19*. Retrieved June 1, 2024, from https://articulo19.org/periodistasasesinados/

Ávila Vallecillo, J. A. (2021). Inteligencia artificial: Discusiones e implicaciones actuales en materia de Derechos de Autor. *Revista de la Facultad de Derecho de México, 71*(281–1), 55–80. https://doi.org/10.22201/fder.24488933e.2021.281-1.80288

Barber, G. (2023, December 7). The Generative AI Copyright Fight Is Just Getting Started. *Wired*. Retrieved March 16, 2024, from https://www.wired.com/story/livewired-generative-ai-copyright/

Beauregard, L. P. (2024, January 24). 'Los Angeles Times' despide a 115 periodistas ante "otro año de fuertes pérdidas." *El País*. Retrieved March 17, 2024, from https://elpais.com/sociedad/2024-01-24/los-angeles-times-despide-a-115-periodistas-ante-otro-ano-de-fuertes-perdidas.html

Beckett, C. (2019). *New Powers, New Responsibilities. A Global Survey of Journalism and Artificial Intelligence.* LSE—Google. Retrieved March 16, 2024, from https://www.journalismai.info/research/2019-new-powers-new-responsibilities

Bedingfield, W. (2023, July 25). IA vs derechos de autor: ¿una batalla perdida? *WIRED.* Retrieved March 16, 2024, from https://es.wired.com/articulos/ia-vs-derechos-de-autor-una-batalla-perdida

Blanco-Herrero, D., Oller Alonso, M., Arcila Calderón, C., Blanco-Herrero, D., Oller Alonso, M., & Arcila Calderón, C. (2020). Working Conditions of Ibero-American Journalists. Temporary and Geographical Differences in Brazil, Mexico, Chile, Spain and Portugal. *Comunicación y Sociedad, 17*, 1–38. https://doi.org/10.32870/cys.v2020.7636

Carvajal, M. (2020, May 11). El impacto del COVID-19 en la economía del periodismo. *Revista de Innovación en Periodismo.* Retrieved March 16, 2024, from https://mip.umh.es/blog/2020/05/11/impacto-covid-19-economia-periodismo/

Castellanos, D. S. (2024, April 12). Uso de TikTok en LatAm supera promedio mundial y es una de las apps más usadas entre la generación Z. *Bloomberg.* Retrieved October 22, 2024, from https://www.bloomberglinea.com/2024/04/12/uso-de-tiktok-en-latam-supera-promedio-mundial-y-es-una-de-las-apps-mas-usadas-entre-la-generacion-z/

Castells, P., & Rattel, F. (2024). *Uso de redes móviles en América Latina, tráfico de datos en la actualidad y proyecciones a 2030.* GSMA Latin America. Retrieved October 31, 2024, from https://www.gsma.com/about-us/regions/latin-america/wp-content/uploads/2024/10/Uso-de-redes-moviles-en-America-Latina-SPA.pdf

Charon, J.-M. (2010). De la presse imprimée à la presse numérique. Le débat français. *Réseaux, 160–161*(2–3), 255–281. https://doi.org/10.3917/res.160.0255

CLIP/IMS. (2021). *Uso de Inteligencia artificial en los medios de comunicación de América Latina.* Centro Latinoamericano de Investigación Periodística (CLIP) e International Media Support (IMS).

Darcy, O., & Passantino, J. (2024, January 25). News industry off to brutal 2024 start as mass layoffs devastate publishers, raising questions about the future of journalism | CNN Business. *CNN.* Retrieved June 2, 2024, from https://www.cnn.com/2024/01/25/media/news-industry-future/index.html

Data México. (2024). *Periodistas y Redactores: Salarios, diversidad, industrias e informalidad laboral.* Data México. Retrieved March 16, 2024, from https://www.economia.gob.mx/datamexico/es/profile/occupation/periodistas-y-redactores

Dave, P. (2021, November 18). Google signs 5-year deal to pay for news from AFP. *Reuters.* https://www.reuters.com/technology/google-signs-5-year-deal-pay-news-afp-2021-11-17/

Díaz Charquero, P., & Rangel, V. (2023). *Informe sobre Políticas de Inteligencia Artificial y Derechos de Autor en América Latina*. Fundación Karisma. Retrieved March 16, 2024, from https://web.karisma.org.co/informe-sobre-politicas-de-inteligencia-artificial-y-derechos-de-autor-en-america-latina/

Editorial La República. (2018, June 21). Al menos 60 medios de comunicación han cerrado durante la era del chavismo. *Diario La República*. Retrieved March 16, 2024, from https://www.larepublica.co/especiales/venezuela-en-picada/al-menos-60-medios-de-comunicacion-han-cerrado-durante-la-era-del-chavismo-2740815

El País. (2005, March 22). Google News retira los contenidos de AFP tras una denuncia de la agencia. *El País*. Madrid. Retrieved March 16, 2024, from https://elpais.com/tecnologia/2005/03/22/actualidad/1111483680_850215.html

Europa Press. (2023, April 12). Universal pide a Apple y Spotify frenar la música generada por IA por no respetar el copyright. *El Economista*. Retrieved October 22, 2024, from https://www.eleconomista.com.mx/tecnologia/Universal-pide-a-Apple-y-Spotify-frenar-la-musica-generada-por-IA-por-no-respetar-el-copyright-20230412-0039.html

FENAJ. (2023). *Panorama da Mobilização e Negociação Salarial nos Sindicatos de Jornalistas*. Federação Nacional dos Jornalistas. Retrieved June 1, 2024, from https://fenaj.org.br/wp-content/uploads/2023/09/PANORAMA-SINDICAL-FENAJ-AGOSTO-DE-2023.pdf

Fletcher, R., Robertson, C. T., Ross Arguedas, A., & Nielsen, R. K. (2024). *Reuters Institute Digital News Report*. Reuters Institute for the Study of Journalism. https://doi.org/10.60625/RISJ-VY6N-4V57

FRANCE 24. (2023, July 1). ¿Por qué están cerrando medios en América Latina? *FRANCE 24 Español*. Retrieved March 16, 2024, from https://www.youtube.com/watch?v=D-fBx3rMqws

Golding, Y. T. (2023, October 18). The news media and AI: A new front in copyright law. *Columbia Journalism Review*. Retrieved March 16, 2024, from https://www.cjr.org/business_of_news/data-scraping-ai-litigation-lawsuit-artists-authors.php

Gollmitzer, M. (2024). Care(lessness) in Precarious Journalism, Before and During the Pandemic: Freelancers' Work-Life Experiences and Coping Strategies. *Global Media and China, 9*(2), 20594364241230434. https://doi.org/10.1177/20594364241230434

González, F. (2023, March 24). Se llama Nat y es la primera presentadora de noticias desarrollada en América Latina con Inteligencia Artificial. *WIRED*. Retrieved October 22, 2024, from https://es.wired.com/articulos/inteligencia-artificial-se-llama-nat-y-es-la-primera-presentadora-de-noticias-desarrollada-en-america-latina-con-ia

Guttmann, A. (2024). *Podcast Consumption in Latin America*. Statista. Retrieved October 31, 2024, from https://www.statista.com/topics/9770/podcast-consumption-in-latin-america/

Herrscher, R. (2017, May 13). La miseria del mejor oficio del mundo. *The New York Times*. Retrieved March 16, 2024, from https://www.nytimes.com/es/2017/05/13/espanol/opinion/la-miseria-del-mejor-oficio-del-mundo.html

Iordanidou, S., Jebril, N., & Takas, E. (Eds.). (2022). *Journalism and Digital Content in Emerging Media Markets*. Springer International Publishing. https://doi.org/10.1007/978-3-031-04552-3

Israde, Y. (2024, February 18). Piden regular uso de IA en la industria editorial. *Mural*. Retrieved February 25, 2024, from https://www.mural.com.mx/piden-regular-uso-de-ia-en-la-industria-editorial/ar2759619

Juárez, B. (2023, July 27). El periodismo en México se escribe entre violencia y precariedad laboral. *El Economista*. Retrieved March 16, 2024, from https://www.eleconomista.com.mx/capitalhumano/El-periodismo-en-Mexico-se-escribe-entre-violencia-y-precariedad-laboral-20230726-0064.html

Kahn, G. (2024, January). *A Shrinking Path: Ten Young Journalists Open Up about their Struggles to Break into the News Industry | Reuters Institute for the Study of Journalism*. Reuters Institute for the Study of Journalism. Retrieved November 5, 2024, from https://reutersinstitute.politics.ox.ac.uk/news/shrinking-path-ten-young-journalists-open-about-their-struggles-break-news-industry

Kaiser, R. (2014). *The Bad News About the News*. The Brookings Institution. Retrieved May 31, 2024, from http://www.brookings.edu/research/essays/2014/bad-news

Langley, A. (2012). Europe's newspapers are dying too. *Columbia Journalism Review*. Retrieved May 31, 2024, from https://www.cjr.org/behind_the_news/european_newspapers_in_dire_st.php

LatAm Intersect PR. (2023, December 19). State of Journalism in Latin America delves into the impact of AI social media and hybrid working styles in 2023. *Latin America PR Agency—LatAm Intersect PR*. Retrieved March 16, 2024, from https://latamintersectpr.com/state-of-journalism-in-latin-america-delves-into-the-impact-of-ai-social-media-and-hybrid-working-styles-in-2023/

LeCompte, C. (2015). *Automation in the Newsroom*. Washington Post: Nieman Reports. Retrieved June 2, 2024, from https://niemanreports.org/articles/automation-in-the-newsroom/

Lee, J.-A., Hilty, R., & Liu, K.-C. (Eds.). (2021). *Artificial Intelligence and Intellectual Property* (1st ed.). Oxford University Press. https://doi.org/10.1093/oso/9780198870944.001.0001

Magnani, L. B. (2024, February 13). Direitos autorais e inteligência artificial. *JOTA Jornalismo*. Retrieved November 4, 2024, from https://www.jota.info/artigos/direitos-autorais-e-inteligencia-artificial

Marconi, F. (2020). *Newsmakers: Artificial Intelligence and the Future of Journalism*. Columbia University Press.

McCormick, S. (2023, May 3). History of the digital newsroom. *Big Bite*. Retrieved May 29, 2024, from https://bigbite.net/2023/05/03/history-of-the-digital-newsroom/

Mioli, T. (2021, August 18). A más de un año del golpe por la pandemia, pequeños diarios de América Latina siguen en recuperación. *LatAm Journalism Review*. Retrieved March 16, 2024, from https://latamjournalismreview.org/es/articles/a-mas-de-un-ano-del-golpe-por-la-pandemia-pequenos-diarios-de-america-latina-siguen-en-recuperacion/

Molina, F. R., & Centenera, M. (2018, July 5). Argentina protesta por el despido de 354 trabajadores de la agencia estatal de noticias. *El País*. Madrid. Retrieved March 17, 2024, from https://elpais.com/internacional/2018/07/05/argentina/1530818310_666821.html

Mullin, B., & Robertson, K. (2024, January 18). Billionaires Wanted to Save the News Industry. They're Losing a Fortune. *The New York Times*. Retrieved January 19, 2024, from https://www.nytimes.com/2024/01/18/business/media/billionaires-news-media-owners.html

Muse Abernathy, P. (2023). *The State of Local News 2023*. Northwestern University. Retrieved May 31, 2024, from https://localnewsinitiative.northwestern.edu/projects/state-of-local-news/2023/report/

Navarro, J. G. (2024). *Digital Advertising in Latin America*. Statista. Retrieved November 5, 2024, from https://www.statista.com/topics/7021/digital-advertising-in-latin-america/

New Media Alliance. (2023). *White Paper: How the Pervasive Copying of Expressive Works to Train and Fuel Generative Artificial Intelligence Systems is Copyright Infringement and Not a Fair Use*. New Media Alliance.

Newman, N. (2023). *Overview and Key Findings of the 2023 Digital News Report*. Reuters/Oxford. Retrieved May 29, 2024, from http://reutersinstitute.politics.ox.ac.uk/digital-news-report/2023/dnr-executive-summary

Paludi, D. P. (2024, September 9). El Futuro de la Propiedad Intelectual: ¿Quién es el Autor en la Era de la Inteligencia Artificial? *Blog Grupo Professional*. Retrieved November 4, 2024, from https://www.grupoprofessional.com.ar/blog/el-futuro-de-la-propiedad-intelectual-quien-es-el-autor-en-la-era-de-la-inteligencia-artificial/

Pennacchio, K. (2023, February 20). Periodistas latinoamericanos denuncian precariedad en la profesión. *LatAm Journalism Review*. Retrieved March 16,

2024, from https://latamjournalismreview.org/es/articles/periodistas-latinoamericanos-denuncian-precariedad-en-la-profesion/

Redacción AN. (2012, November 10). 'El País' despide a 129 periodistas… vía correo electrónico. *Aristegui Noticias*. Retrieved March 17, 2024, from https://aristeguinoticias.com/1011/mundo/el-pais-despide-a-129-periodistas-via-correo-electronico/

Rinehart, A., & Kung, E. (2022). *Artificial Intelligence in Local News*. The Associated Press.

Rivera Escobedo, A. (2020). *"Ahora somos prácticamente esclavos": La precariedad laboral del periodismo en México* (Maestría en Periodismo sobre Políticas Públicas). CIDE, Ciudad de México. Retrieved from http://repositorio-digital.cide.edu//handle/11651/4251

Salaverría, R., & de-Lima-Santos, M.-F. (Eds.). (2021). *Journalism, Data and Technology in Latin America*. Springer International Publishing. https://doi.org/10.1007/978-3-030-65860-1

Salaverría, R., Sádaba, C., Breiner, J. G., & Warner, J. C. (2019). A Brave New Digital Journalism in Latin America. In M. Túñez-López, V.-A. Martínez-Fernández, X. López-García, X. Rúas-Araújo, & F. Campos-Freire (Eds.), *Communication: Innovation & Quality* (pp. 229–247). Springer International Publishing. https://doi.org/10.1007/978-3-319-91860-0_14

Sanjurjo, S., & de Frutos, R. (2022). Impacto del COVID-19 en el periodismo latinoamericano: Entre la precariedad laboral y las secuelas psicológicas. *Cuadernos.info, 51*, 114–137. https://doi.org/10.7764/cdi.51.27329

Sarmiento, J. (2024, May 2). ANP: Periodistas ganan muy poco y trabajan en precarias condiciones. *Wayka.pe*. Retrieved June 1, 2024, from https://wayka.pe/anp-periodistas-ganan-muy-poco-y-trabajan-en-precarias-condiciones/

Simonetti, I. (2022, June 29). Over 360 newspapers have closed since just before the start of the pandemic. *The New York Times*. Retrieved March 16, 2024, from https://www.nytimes.com/2022/06/29/business/media/local-newspapers-pandemic.html

Sjøvaag, H. (2024). The Business of News in the AI Economy. *AI Magazine, 45*(2), 246–255. https://doi.org/10.1002/aaai.12172

The Economist. (2024, April). Generative AI is a marvel. Is it also built on theft? *The Economist*. https://www.economist.com/business/2024/04/14/generative-ai-is-a-marvel-is-it-also-built-on-theft

Wachowicz, M. (2019). *Inteligência artificial e criatividade: Novos conceitos na propriedade intelectual*. Gedai—Ufpr.

Watson, A. (2024). *Printed Media in Europe*. Statista Research Department. Retrieved May 31, 2024, from https://www.statista.com/topics/4031/printed-media-in-europe/

Wilding, D. (2018). *The Impact of Digital Platforms on News and Journalistic Content*. The Centre for Media Transition.

Wölker, A., & Powell, T. E. (2021). Algorithms in the Newsroom? News Readers' Perceived Credibility and Selection of Automated Journalism. *Journalism, 22*(1), 86–103. https://doi.org/10.1177/1464884918757072

Zuazo, N. (2023). *Periodismo e inteligencia artificial en América Latina*. UNESCO Biblioteca Digital. Retrieved March 16, 2024, from https://unesdoc.unesco.org/ark:/48223/pf0000388124

From Epistemological Foundations over Futuristic Speculations to Fact-Based Concerns: Evolving Discussions on Artificial Intelligence in Argentina, Colombia, and Mexico (2014–2024)

Fátima Ávila-Acosta, Lucía Morales-Lizárraga, and Jan Nehring

Introduction

Artificial Intelligence (AI) is a critical protagonist in the current public debate. A spectrum of attitudes, from optimism to skepticism, has been raised around AI. While some embrace its transformative potential, others express concerns about its accelerated development, highlighting its ethical, social, and economic impacts, and raising elaborate narratives around

F. Ávila-Acosta (✉)
Berlin Graduate School of Social Sciences, Humboldt University of Berlin, Berlin, Germany
e-mail: acostafa@hu-berlin.de

D. Ramírez Plascencia, R. M. Alonzo González (eds.), *Ethical and Legal Dilemmas of Artificial Intelligence in Latin America*,
https://doi.org/10.1007/978-3-031-86540-4_3

AI's uncertainties. As AI continues to integrate into daily life and expand across industries, understanding public opinion toward it remains essential, as it shapes societal acceptance, influences policy, and frames ethical considerations for AI governance.

Existing literature on AI attitudes has primarily focused on Western countries (Fast & Horvitz, 2016; Chuan et al., 2019; Ouchchy et al., 2020; Kelley et al., 2021) with surveys, focus groups, and media and social media analyses offering a window to explore public perceptions on AI (Nader et al., 2022; Kelley et al., 2021; Garvey & Maskal, 2019). Analyses of Western countries highlight a general optimism surrounding AI, tempered by concerns about its impact and a noticeable polarization of the discourse with varying degrees of emphasis on the benefits and risks of AI. However, research examining public attitudes toward AI in Latin America remains limited, with little exploration of how the region's cultural, economic, and political contexts shape local perceptions and discussions on the technology.

This chapter explores this gap by understanding AI's reception in non-Western societies, specifically in Latin America. We examine the prominent themes, trends, and sentiments that shape the public debate on AI in Mexico, Argentina, and Colombia from 2014 to 2024. By analyzing opinion columns from two major media outlets per country, this study aims to answer: What are the predominant topics, sentiments, and opinions associated with AI as framed by influential voices in the media in these countries? Although the sample may not fully represent all the diverse voices and groups in the region, opinion columns provide access to subjective and expressive views, providing insights into the narratives and beliefs of influential voices within Latin American society. Through this focus, we aim to explore and understand the evolving dynamics of public opinion on AI in Latin America and its broader implications.

L. Morales-Lizárraga
University of Guadalajara, Guadalajara, Jalisco, Mexico
e-mail: lucia.morales@udgvirtual.udg.mx

J. Nehring
Centre for the Fourth Industrial Revolution Rwanda (C4IR), Kigali, Rwanda
e-mail: jnehring@c4ir.rw

Background

Public perceptions are crucial in shaping societal and policy dynamics by influencing which topics are prioritized on public and policy agendas. Kingdon's (1995) agenda-setting theory highlights how public opinion draws attention to specific issues, framing them as matters of societal importance. Similarly, McCombs and Shaw's (1972) media agenda-setting theory demonstrates how media coverage shapes public perceptions, ultimately guiding policy priorities. Baumgartner and Jones (2009) further show that shifts in public sentiment can pressure governments to enact policy changes. In the context of innovations like AI, public opinion is vital for shaping societal acceptance and driving regulatory and ethical debates (Fast & Horvitz, 2016; Ouchchy et al., 2020). Understanding public perceptions thus provides critical insights into how new topics gain prominence and influence policy responses.

Building on this, researchers have sought to understand how societies perceive AI, given its growing prominence in public and policy discussions. Some approaches include the study of the public understanding of AI through surveys and focus groups (Nader et al., 2022; Kelley et al., 2021). Another important part of research has focused on analyzing social media and news coverage. Following this perspective, Fast and Horvitz (2016) analyzed the New York Times' coverage of AI over 30 years (1986–2016), finding a predominantly optimistic tone alongside growing concerns about the societal impacts of AI. Similarly, Garvey and Maskal (2019) analyzed media coverage in the United States over six decades (1956–2018), revealing no predominant negative sentiment toward AI, contrary to common assumptions. Moriniello et al. (2024) conducted a sentiment analysis of AI coverage in WIRED magazine from 2018 to 2023. They observed an overall positive sentiment with an increasing polarization in both positive and negative viewpoints over time.

Chuan et al. (2019) performed a content analysis of US newspaper coverage of AI between 2009 and 2018, finding that the dominant topics were business and technology. Their study also highlighted that while the benefits of AI were discussed more often, the risks were explored in greater detail. Ouchchy et al. (2020) analyzed media portrayals of AI's ethical issues in English-language sources, finding that the coverage was generally realistic and practical but lacked depth, calling for a more comprehensive exploration of these issues.

While most related studies on AI and public perceptions have predominantly focused on English-speaking countries, there have been increasing efforts to expand this focus to other regions. For example, Canavilhas et al. (2024) analyzed how AI is represented in Brazilian and Portuguese media from 2022 to 2024, finding predominantly positive coverage of AI and focusing on its potential benefits. Similarly, Hernandes and Corsi (2024) analyzed the Brazilian media from 2023 to 2024 and found a strong focus on AI applications in the workplace and product launches, with visible corporate influence over the media representation of AI. In Spanish-speaking countries, Parratt et al. (2024) studied media coverage in Spain, revealing that whereas the presence of AI is on the rise, news reports focus on issues related to automation without significant emphasis on job loss or ethics.

Despite these important contributions, much of the existing literature concentrates on English-speaking countries, particularly the USA. The perception of AI in Latin America, with its different cultural, social, and political contexts, remains understudied. This chapter addresses this gap by examining the public opinions on AI in three Latin American countries—Argentina, Colombia, and Mexico—through the lens of opinion columns in prominent media outlets. Additionally, this study takes a comparative approach, analyzing AI-related public discourse in the three studied countries and contributing to the broader regional conversation about the evolving views on AI.

Case Selection

For our study, we focus on Argentina, Colombia, and Mexico. These three countries are among the top economies in Latin America, playing a central role in regional political, social, and economic dynamics, including technological developments. In addition, these countries have diverse and influential media outlets, playing a critical role in framing narratives on different public topics. At the same time, the varied socio-political contexts of each country make their comparison interesting and enrich the discussion in the study of regional variations in AI coverage and public opinion.

In the case of Argentina, the country began designing a National AI Plan in 2018 as part of its Digital Agenda 2030 to provide a framework for the technology's development. However, the plan was never finalized and remains in a consultative status. In 2022, the country approved the

"Science, Technology and Information Plan 2030." The country has participated in international discussions on AI ethics and adopted UNESCO's recommendations on the subject. Recently, Argentina's controversial president, Javier Milei, has expressed ambitions to position the country as a global AI Hub, meeting with figures like Mark Zuckerberg and Elon Musk to discuss innovation. In August, Milei established the "Artificial Intelligence Applied to Security Unit," tasked with monitoring social media to predict and prevent crimes. This initiative has sparked significant debate, as human rights organizations raise concerns over privacy, surveillance, and potential misuse of such technologies.

Colombia has made significant strides in digital transformation and AI policy. In 2019, the National Policy for Digital Transformation and AI established ethical principles to guide AI development. This was followed by the National Digital Strategy in 2023, which prioritized the responsible adoption, implementation, and democratization of AI. Building on these efforts, Bill 447 was introduced in 2024 to promote responsible data usage for public policy and AI production, emphasizing privacy, as presented by the Minister of Information and Communication Technologies. The judicial system has also actively engaged with UNESCO to ensure the ethical use of AI.

In Mexico, efforts to advance AI began with an initial governmental push in 2018, marked by active participation in international debates and regional initiatives. Although the country currently lacks a national AI strategy, Mexico has made notable progress in regulating AI-related issues, particularly in privacy, data protection, and the protection of human rights, including the criminalization of AI misuse. Key challenges remain, such as developing mechanisms for societal inclusion in AI governance and regulatory frameworks to address socio-economic inequalities, gender gaps, and the limited integration of Indigenous languages in AI systems (UNESCO, 2022).

Newspaper Selection

With their expressive and subjective tone, opinion articles offer valuable insights into public sentiment and societal debates. They reflect and influence public opinion by articulating personal viewpoints. Jacobs and Townsley (2011) highlight their importance in shaping public perception, noting their key role in media, democracy, and the public sphere.

Building on this understanding, we selected opinion articles from prominent online newspapers to analyze public discourse on AI. For the newspaper selection, we relied on the Reuters Institute Digital News Report 2023, specifically focusing on the "Top Brands of Media" listed in the report (Newman et al., 2023). Moreover, we used only newspapers with appropriate search functionalities to assist data collection. We selected the following newspapers: for Argentina, La Nación and Clarín; for Colombia, El Espectador and Semana; and for Mexico, Milenio and El Universal.

Each country showcases unique media landscapes influenced by historical, political, and cultural factors. In Argentina, the media is dominated by private ownership, with a declining influence of public broadcasters and reduced interest in news since the COVID-19 pandemic (Newman et al., 2023). Clarín, a historically influential outlet, and La Nación, known for its conservative-liberal stance, reflect the country's polarized political environment, often critical of current leadership (Becerra & Mastrini, 2017; Newman et al., 2023). In Colombia, Semana magazine and El Espectador newspaper play central roles in shaping political discourse. Semana is renowned for its investigative journalism, particularly on corruption and public affairs, while El Espectador emphasizes editorial independence and supports peace initiatives, maintaining a tradition of center-left to centrist reporting (Rey, 2012; Waisbord, 2013; Llorente & Abad, 2020; Newman et al., 2023). El Universal and Milenio anchor Mexico's media landscape. El Universal, often regarded as a national record newspaper, and Milenio noted for its occasional alignment with government positions, offer comprehensive coverage while reflecting subtle editorial differences (Lawson, 2002; Hughes & Márquez, 2017).

Methodology

For the analysis of the opinion articles in the Argentinean, Colombian, and Mexican media, we programmed a Python script to download all articles in the opinion section from each of the selected online newspapers from January 2014 until the first week of December 2024. Opinion columns had to contain at least one mention of the keywords "*inteligencia artificial*" or "*inteligencias artificiales*" (artificial intelligence, singular and plural).

This initial collection included 6268 opinion articles. However, after reviewing random samples, we found that some articles only mentioned AI in passing, without meaningfully engaging in discussion or debate.

Table 1 Distribution of articles by country and newspaper

Country	*Newspaper*	*Articles*
Argentina	Clarín	62
	La Nación	296
Colombia	El Espectador	44
	Semana	58
Mexico	El Universal	376
	Milenio	151

Therefore, to ensure substantial engagement with the topic of AI, we filtered for articles with at least two mentions of the keywords, which reduced the dataset to 2352 articles. After identifying and removing duplicate entries, the final dataset consisted of 987 unique articles, distributed as follows: 358 from Argentina, 102 from Colombia, and 527 from Mexico. Table 1 shows the number of articles per newspaper.

Given the large number of articles, we chose automatic analysis methods. We conducted a comparative frequency analysis, sentiment analysis, opinion identification, and opinion classification.

Text Analysis with Large Language Models

Large language models (LLMs) provide state-of-the-art results in various natural language processing tasks (OpenAI, 2023), such as sentiment classification (Zhang et al., 2024) and opinion mining (Ahmadnia et al., 2024). While traditional methods, such as fine-tuned transformers (Devlin et al., 2019), often perform better in scenarios with specific training data, LLMs excel in situations with little or no training data (Brown et al., 2020). This makes LLMs ideal for our analysis, as we lack training data.

For our analysis, we used GPT-4o (version gpt-4o-2024-08-06), a multilingual LLM by OpenAI capable of processing Spanish and English with high quality. Currently, it is OpenAI's most advanced model (OpenAI, 2025). To conduct the sentiment analysis and the opinion identification and classification, we described the task in a short text in natural language (also known as "prompt") to the LLM and provided a few-shot examples to improve its performance (Brown et al., 2020).

Sentiment Analysis

Sentiment analysis involves collecting and examining people's opinions, thoughts, and feelings about different topics, products, subjects, or services (Wankhade et al., 2022). Textual documents usually contain a variety of sentiments toward diverse topics. Many sentiment tools only provide a general sentiment toward all topics. In contrast, our LLM-based approach classifies the sentiment toward AI, providing more decisive results than general sentiment.

For our analysis, we assigned positive, neutral, and negative labels to the opinion columns considering their views on AI. Positive sentiment reflects enthusiasm or a favorable attitude toward AI, highlighting its benefits, advancements, and positive impact on society. Neutral sentiment presents balanced, objective information, discussing pros and cons without strong emotional judgments. Negative sentiment expresses concerns, skepticism, or criticism, focusing on risks, ethical issues, and potential adverse effects of AI on, e.g., employment, privacy, and society.

Opinion Identification and Classification

Opinion identification and classification involve detecting and categorizing opinions expressed in text. The LLM analyzes the content to identify statements of opinion and then classify them into specific categories. We conducted opinion identification and classification for a more targeted analysis of public sentiment on AI. We submitted three random samples of 97, 101, and 108 articles to GPT-4o, considering token limits, with a prompt asking it to identify the 10 most important AI-related topics. After combining the 30 identified topics, we manually reviewed and grouped similar ones, resulting in three main categories: *Enthusiasm*, *Skepticism and Caution*, and *Fear and Concern*.

Enthusiasm reflects positive emotions about AI, e.g., its potential to drive innovation and improve many significant aspects of life. *Skepticism and Caution* involve doubts and ethical concerns about AI, focusing on risks, overblown promises, and the call for responsible use and regulation. *Fear and Concern* highlight anxieties related to the adverse effects of AI, such as job loss, misinformation, and lack of control, coupled with frustration over inadequate regulation and concerns about AI's rapid development. After defining these categories, we asked GPT-4o to categorize opinions in each article. We provided a definition and examples for each category in the prompt, allowing every article to have more than one category if that was the case.

Comparative Frequency Analysis

Comparative frequency analysis (CFA) counts how often specific keywords or phrases appear in a text collection and compares these counts across different groups or categories. It identifies terms that stand out in one group (e.g., articles from 2019) compared to a baseline group (all other articles). CFA assigns higher values to terms common in the target group but rare in the baseline. The formula for CFA is shown below:

$$\text{Term\_frequency}\left(\text{term, collection}\right) = \text{Number of occurrences of term in the collection} \tag{1}$$

$$\text{Comparative\_term\_frequency}\left(\text{term}\right) = \frac{\text{term\_frequency}\left(\text{term, target corpus}\right)}{\text{term\_frequency}\left(\text{term, baseline corpus}\right)} \tag{2}$$

We excluded terms that occur less than three times in either of the groups. Further, we ranked the terms by CFA and selected the top 30 for each analysis. This was done for articles grouped by year and country, as well as both country and year. From the top 30 prevalent terms for each corpus, we carefully reviewed and retained only those terms relevant to our analysis based on their association or connection to AI debates.

LLM Methods Evaluation

To evaluate the performance (prediction quality) of the LLM methods, we manually annotated a random sample of 10% of the dataset's articles using the same categories employed for sentiment analysis and opinion classification. Articles were annotated on a binary scale (yes/no) to indicate whether the sentiment or opinion classification was present. We compared the LLM predictions with our human annotations to measure precision, recall, and F1-score, as well as assess the performance of sentiment analysis and opinion classification. Additionally, we annotated 24 samples with two annotators to compute inter-annotator agreement (Cohen's kappa) to assess the quality of the human labels. Further, Support represents the number of samples in our gold standard. The following table summarizes the evaluation metrics.

As shown in Table 2, the sentiment analysis has a Cohen's kappa value of 0.85, indicating almost perfect agreement, with F1-scores of 0.56 for

Table 2 Evaluation metrics for sentiment analysis and opinion classification

Category	*Cohen's kappa*	*Precision*	*Recall*	*F1-score*	*Support*
Sentiment analysis					
Positive	0.85 (almost perfect)	1.00	0.39	0.56	46
Neutral		0.40	0.73	0.51	26
Negative		0.78	0.96	0.86	26
Weighted average		**0.78**	**0.63**	**0.63**	**98**
Opinion classification					
Enthusiasm	0.75 (substantial)	0.83	0.83	0.83	52
Fear and concern	0.83 (almost perfect)	0.52	1.00	0.69	22
Skepticism and caution	0.38 (fair)	0.61	0.86	0.72	44
Weighted average	–	**0.69**	**0.87**	**0.76**	**132**

Positive, 0.51 for Neutral, and 0.86 for Negative, resulting in a weighted average of 0.63, suggesting satisfactory classification quality. In opinion classification, Cohen's kappa values show mixed results: *Enthusiasm* (0.75, substantial), *Fear and Concern* (0.83, almost perfect), and *Skepticism and Caution* (0.38, fair). The F1-scores for opinion categories are *Enthusiasm* 0.83, *Fear and Concern* 0.69, *Skepticism and Caution* 0.72, with a weighted average of 0.74. While the inter-annotator agreement ranges from fair to almost perfect, the quality of opinion classification is generally acceptable to good across all classes. The performance of the AI methods is good to be used in our analysis.

Results

General Trends and Distribution of AI-Related Opinion Columns

The following figures present the overall distribution of opinion columns engaging in AI-related discussions. Specifically, Fig. 1 displays the yearly distribution of articles aggregated by all countries, while Fig. 2 shows the yearly distribution by country.

Overall, the data reveals a positive trend in the publication of AI-related content over the analyzed period. Coverage was relatively low in the early years but consistently grew over time. However, there was a slight decline in AI-related publications between 2020 and 2022, followed by a particularly sharp rise after 2022.

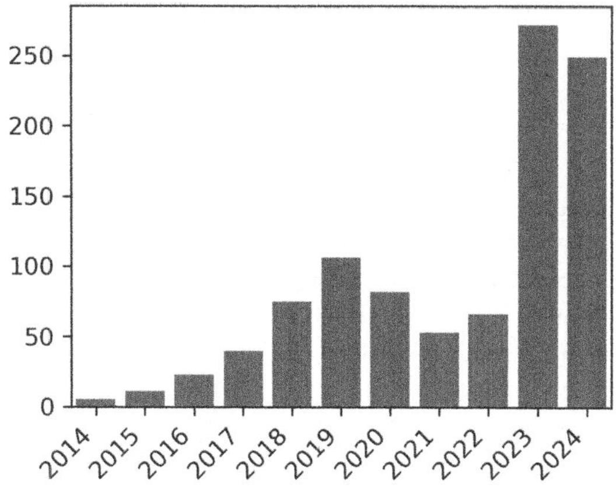

Fig. 1 Number of articles per year

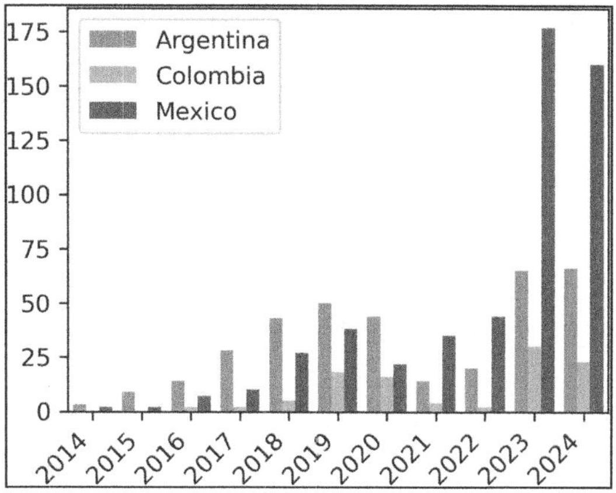

Fig. 2 Number of articles per year and country

When analyzing the yearly distribution of articles by country, this general upward trend is evident across all three countries. Between 2014 and 2020, Argentinian newspapers led the coverage of AI-related opinion columns. However, from 2021 onward, Mexico emerged as the leader in publishing AI-focused opinion pieces. Notably, all three countries peaked in AI-related opinion column publications in 2023, aligning with the aggregated trend. In particular, Mexico registered a significantly higher number of opinion columns than the other countries in 2023 and 2024. The numbers suggest that fewer articles were published in 2024 compared to 2023, but this comparison is not valid because we stopped data collection during the first half of December 2024.

Sentiment Analysis

Figures 3 and 4 present the aggregated distribution of sentiments per year and country.

Neutral sentiment toward AI has been the most common over the years, except for 2023 and 2024, when negative sentiment became the dominant view. Apart from 2019, negative opinions on AI have consistently ranked as the second most common sentiment, showing a significant increase in 2023 and 2024, ultimately becoming the majority view.

Regarding the aggregated distribution by country (Fig. 4), the general trend of predominantly neutral sentiments, followed by negative ones, is consistent across all three countries. However, there is a notable difference in positive sentiments. While the percentage of neutral and negative articles is similar between countries, Mexico stands out with a significantly higher proportion of positive opinion columns about AI compared to Argentina and Colombia.

Figure 5 presents the annual evolution of sentiment trends across the analyzed countries, highlighting variations in how opinions toward AI shifted over time in each context.

We observe that the trends in Argentina and Colombia align with the general pattern, showing predominantly neutral sentiments, with negative

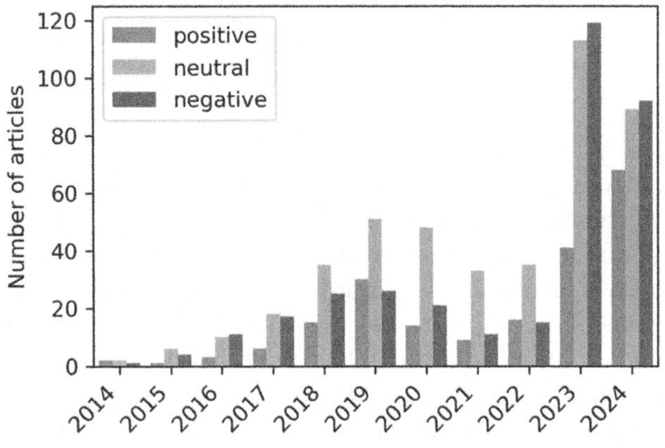

Fig. 3 Yearly distribution of sentiments

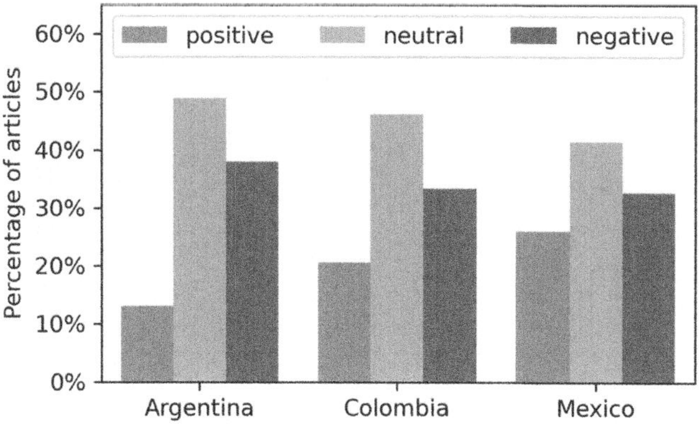

Fig. 4 Distribution of sentiments per country

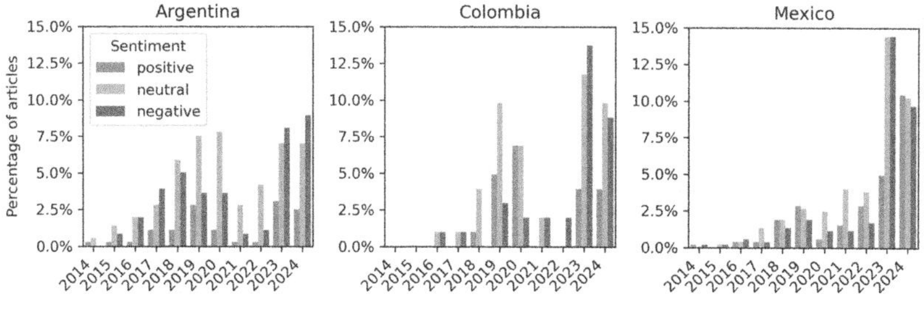

Fig. 5 Distribution of sentiments per year and per country

sentiment becoming dominant in 2023 for Colombia and 2023 and 2024 for Argentina. However, while Mexico follows the general trend, it shows a notable difference in 2023, where the number of negative and neutral articles is equal. By 2024, positive sentiment becomes the leading category in Mexico, slightly surpassing neutral and negative sentiments.

Opinion Classification

Figure 6 shows the distribution of opinions detected in the articles over the 10-year analysis, while Fig. 7 displays the percentage of yearly opinions by country.

The dominant category throughout the analyzed period is *Fear and Concern*, except in 2021 and 2022. Although the differences between

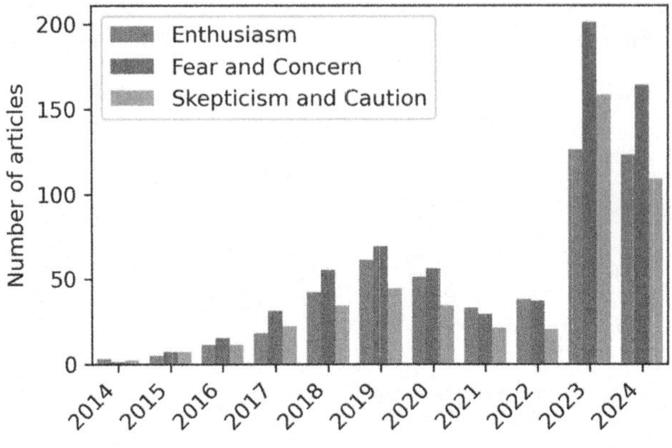

Fig. 6 Yearly distribution of opinions

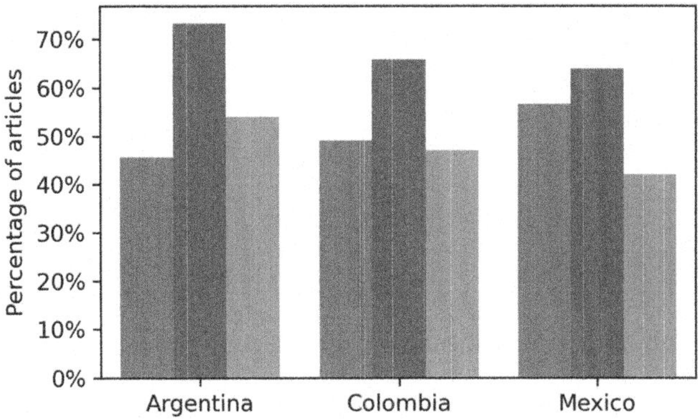

Fig. 7 Distribution of opinions per country

categories are not very pronounced in most years, a notable shift occurs in 2023 and 2024, with a sharp increase in articles expressing *Fear and Concern*. The aggregated data by country (Fig. 7) shows that *Fear and Concern* remains the leading category. Interestingly, Mexico reports the fewest articles in the *Fear and Concern* category but the most with an *Enthusiasm* view of AI compared to the other two countries.

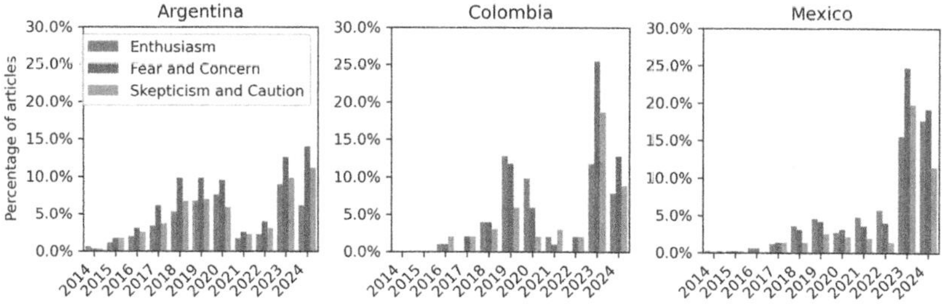

Fig. 8 Distribution of opinions per year and per country

Figure 8 shows the distribution of opinions per year and per country.

In Argentina, the dominant opinion in AI-related articles throughout the analyzed period is consistently *Fear and Concern*. In Colombia, the trends are more varied: for the majority of the period, the leading opinions alternate between *Enthusiasm* and *Skepticism and Caution*. However, in the last two years of analysis, *Fear and Concern* becomes the prevailing perspective. *Enthusiasm* is the leading opinion category in Mexico until 2023, when it is overtaken by *Fear and Concern*, which becomes the most prominent viewpoint.

Comparative Frequency Analysis

Building on the trends identified in the sentiment analysis, the opinion classification further explores the nature of public concerns and enthusiasm about AI. We run a comparative frequency analysis of the terms related to AI retained for our analysis. The following figures present the relevant terms, with their size indicating the CFA of the terms. Figure 9 presents a manual selection of terms from the CFA analysis over the years.

Figure 9 highlights terms such as "computer," "futurist," "GPT," "Turing," "Hawking," and "neuroscience." Less frequent terms include: "epidemic," "AlphaGo," "Frankenstein," and "Coronavirus." In general, these terms describe the range of new emerging technologies, the key players in this arena, and their impacts on society, demonstrating how diverse and constantly changing AI is and how its applications will be.

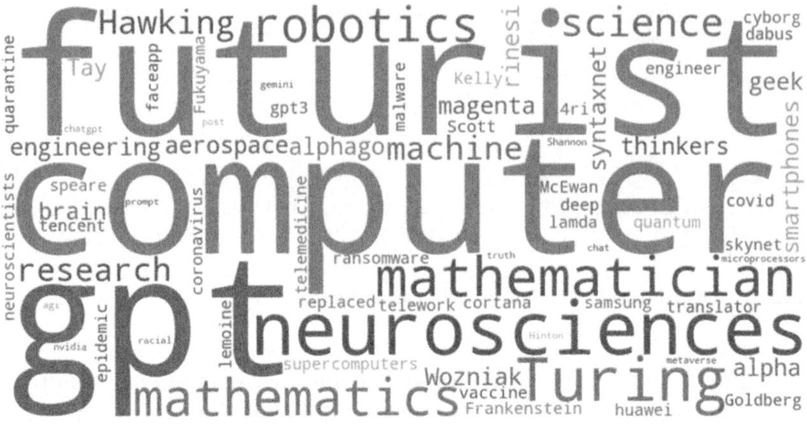

Fig. 9 Aggregated comparative frequency analysis

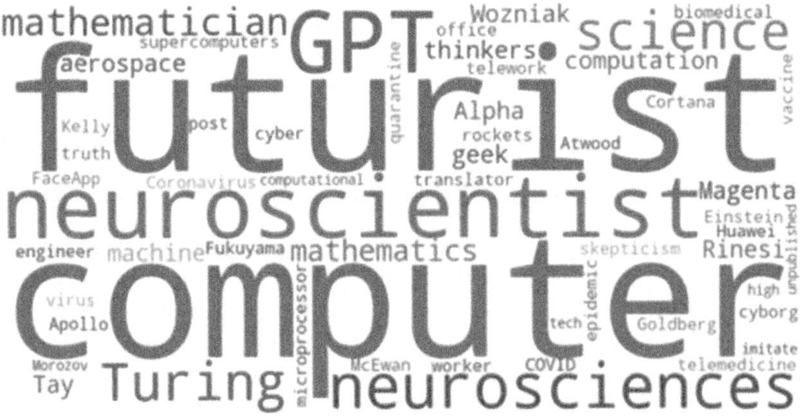

Fig. 10 Argentina: comparative frequency analysis

Figures 10, 11, and 12 illustrate the variation of prominent terms across countries. For instance, in Argentina, notable terms include "computer," "futurist," and "neuroscientist," while in Colombia, terms like "crime," "robot," and "Hawking" stand out. In Mexico, prominent terms include "machine," "robot," and "Deep," referring to Deep Language.

Fig. 11 Colombia: comparative frequency analysis

Fig. 12 Mexico: comparative frequency analysis

Regarding CFA per year and considering the space and word limitations, we developed the following table (Table 3) considering the main yearly trends that emerged from the yearly frequency analysis. The data exhibits clear patterns across the years, which can be divided into distinct periods marked by dominant topics and themes.

Table 3 Aggregated comparative frequency analysis per year

Period	*Name*	*Description*	*Highlights*
2014–2016	Foundational and conceptual stage	Discussions focused on AI's foundational, theoretical, and basic technical concepts, exploring its historical and philosophical origins as well as its early advancements. These included examinations of its initial technical achievements and the emergence of ethical concerns. Broader conversations also addressed AI's future potential and its anticipated impact on society	– Stephen Hawking's and Steve Wozniak's early warnings on AI – AlphaGo's victory over Lee Sedol – Introduction to concepts like neuroscience, robotics, and informatics – Early references to "thinking machines" – AI's potential applications
2017–2018	Expansion, application and integration of AI in society	Discussions centered on AI's practical integration into society and its interaction with humanity, including its blending with human capabilities and the ethical implications of such advancements. AI tools began to gain attention in public discourse, sparking debates about their potential and raising concerns about the risks of possible autonomy	– Mentions of AI's transformative power in society (including the job market and AI's integration with human capabilities) – Discussions on AI's creative capabilities in arts and literature – "Frankenstein-like" concerns about creations turning against their creators
2019–2021	Concerns on AI's accelerating progress	Increased awareness and concerns of the socio-economic, ethical, and security challenges posed by AI. These concerns include ethical and philosophical questions surrounding AI's accelerating progress, while also noting the increasing overlap between science fiction and the real world	– Debates around privacy and cybersecurity – Mentions of AI's role in the pandemic and in transforming the health sector – Comparisons with dystopian sci-fi scenarios – Integration of advanced technologies in society – Mentions of ethics and regulation
2022–2024	Social incorporation and adoption of generative and advanced AI and its impacts	Debates around the adoption of cutting-edge AI language models and their ethical and social implications, particularly the growing concerns of AI-driven misinformation and manipulation. Additional concerns include, algorithmic biases and their potential impact on human rights. There is also an increased focus on the practical integration of AI into daily life, with more specific ethical concerns arising as AI becoming an integral part of everyday activities	– References to Lemoine's case and concerns around AI's potential to develop consciousness and awareness – Mentions of ethics and regulation – Recognition of AI's progress and LLMs popularity – Several references to ChatGPT

Findings and Discussion

Growing AI Presence in Media and Refined Concerns

The inclusion of debates and discussions on AI has increased significantly over the years in all three countries. Over the past decade, AI has transitioned from a niche topic to a central subject in media discourse, which reflects its growing adoption and integration into society. Early discussions were technical and exploratory; some often referred to AI as something out of a science fiction movie, with the idea of its widespread use seeming like a distant future possibility. As time passed, the conversation shifted to focus on the early incorporation of AI in society, raising curiosity about its transformative capabilities. Finally, the discussion evolved to center on the exponentially growing integration of AI in daily life and AI's real-world implications, with increasing urgent concerns about ethical issues, regulation, and its overall socio-economic and governance impacts.

Ethical concerns surrounding AI have been a constant and recurring topic. These ethical concerns evolved from broad and general fears to more specific and well-defined issues over time. In the early years, ethical concerns were focused on abstract worries about AI's potential dangers, such as the risk of surpassing human control. As AI became more integrated into daily life, these concerns became more concrete, addressing very specific and tangible risks such as misinformation, data privacy, algorithmic bias, and the potential impact on human rights. Two global events identified in our analysis—the COVID-19 pandemic and the launch of ChatGPT—further shaped the discourse on AI ethics and its societal implications.

The Effects of the COVID-19 Pandemic and the Introduction of ChatGPT

The COVID-19 pandemic marked a turning point in AI-related discussions, coinciding with a marked decrease in AI-related articles during its most critical years (2020–2022). During this period, attention was turned almost exclusively to pandemic-related issues, as reflected in the surge of COVID-related terms in opinion columns. While neutral sentiments toward AI persisted during the pandemic, we noted that the pandemic played a significant role in accelerating the focus on AI's capabilities, particularly its role in reshaping healthcare and work environments.

Throughout this period, key themes emerged, such as human-AI interaction, automation, and AI's integration into various sectors, particularly healthcare. This trend is underlined by a changed sentiment from *Fear and Concern* in 2020 to more *Enthusiastic* perspectives in 2021 and 2022, likely driven by the hope that technology and AI could play a significant role in alleviating the challenges of the pandemic.

The release of ChatGPT on November 30, 2022, represented another pivotal moment, dramatically increasing the volume of AI-focused opinion columns in 2023 and 2024, and sparking widespread discussions about its ethical, social, cultural, and economic repercussions. ChatGPT's accessibility and its ability to demonstrate the power of generative AI captured the public imagination and drew attention to both its potential and risks, making it a central topic in media discussions. According to our analysis, the launch of ChatGPT marked a sharp rise in negative media sentiment toward AI, along with a peak of articles expressing *Fear and Concern* about AI. This shift to pessimistic views was accompanied by a rise in concerns about the ethics and regulation of AI, along with growing discussions about its implications on areas like labor, education, and governance.

The rapid adoption of ChatGPT in daily tasks and specific economic sectors underscored its disruptive potential, serving as a turning point in public perception of AI which brought justified concerns about the unprecedented pace of AI advancements to life. Its worrisome but remarkable ability to replicate activities—such as writing, creating, and analyzing—that had, until then, been exclusively human, sparked debates and reflections on the transformative power of AI. For many, ChatGPT represented the moment where dystopian views of the future converged with reality, turning speculative fears about AI's integration into daily life into tangible and well-defined concerns.

National Trends and Localized Discussions

In general, individual countries' opinions on AI primarily revolve around *Fear and Concern*, in contrast to the predominance of neutral sentiments toward AI. In the early stages, futuristic perceptions of AI dominated the comparative frequency analysis. Neutral sentiments toward AI prevailed during this period due to the nascent stage of AI adoption. At this stage, AI was perceived more as a distant possibility than an immediate societal force, resulting in balanced, observational tones in media coverage.

However, *Fear and Concern* rose sharply as AI technologies like ChatGPT became more integrated into daily life and began demonstrating tangible impacts on labor, education, and governance. These concerns reflect growing unease about ethical issues, regulatory gaps, and the rapid pace of AI development, which transformed speculative risks into immediate challenges.

Looking closer at the individual countries, we find a clear tendency to frame AI debates within each country's specific context and geographic reality, emphasizing local perspectives, contributions, and implications. Such localized discussions of AI in national and local contexts reflect how societies construct their particular narratives on AI based on their unique cultural, economic, and social contexts.

For Argentina's specific case, its leadership in AI opinion columns until 2020 suggests early engagement with the topic. Argentina's discussions have consistently focused on philosophical and ethical issues and deeper implications of AI, particularly the concept of human consciousness and AI's attempt to replicate human brain functions.

In Colombia, the conversation focused more on the intersection of AI and public safety. This includes debates on how AI can improve security strategies and prevent crime and concerns about its use in facilitating criminal activities, such as cybercrime or fraud. In the early years, Colombian debates focused on workforce automation and AI's integration into industries, raising concerns about its impact on labor.

Mexico's rise as the leader in AI publications from 2021 onward reflects a growing recognition of AI's societal and economic impact, with its media becoming key platforms for discussing the technology's future implications. The Mexican focus has been on cutting-edge technological advancements, practical applications in everyday life, and the expanding capabilities of AI. Over the years, Mexican media has shown a persistent curiosity and fascination with AI, which is reflected in the relatively higher *Enthusiasm* of the opinion columns compared with other countries.

Limitations

Our analysis is based solely on opinion articles from a selection of prominent newspapers, which, although influential, may not fully represent the diversity of public opinion in the region. By focusing on elite media outlets, the study primarily captures the perspectives of influential voices and decision-makers, potentially overlooking grassroots or marginalized viewpoints that might offer alternative insights into AI's societal impact.

Conclusions

Authors have occupied themselves with the question of how societies perceive AI. Existing research has mostly focused on English-speaking countries, with only a few recent studies focused on non-English-speaking countries such as Brazil, Portugal, and Spain. In this chapter, we address this gap by analyzing three key countries in Latin America, shifting the focus of AI perception research to the region, and offering an innovative comparative approach. Using an original database of opinion columns from major media outlets, this chapter analyzed the predominant topics, sentiments, and opinions about AI in Argentina, Colombia, and Mexico over the past 10 years.

We find an increased and growing presence of AI in the media throughout the years, accompanied by a shift in the discourse. Initially, the focus was on theoretical and exploratory discussions, followed by curiosity about AI's early capabilities and gradual adoption. More recently, the discourse highlights AI's undeniable integration into daily life, along with clearly and justified concerns. While ethical concerns have been a recurrent and constant theme throughout the years, these have evolved from broad, often dystopian fears to more specific and localized concerns. These shifts reveal the dual role of media as both a mirror of societal attitudes and an active participant in shaping public opinion and policy agendas.

In our analysis, we detected two major events that significantly shaped the trajectory and acceleration of AI-related discussions: the COVID-19 pandemic and the release of ChatGPT. The effect of these events was felt across all three countries. The pandemic underscored AI's potential to address critical societal challenges, such as reshaping healthcare and work environments, fostering initial *enthusiasm*. However, the launch of ChatGPT in late 2022 marked a turning point, intensifying concerns about misinformation, ethical regulation, and the broader societal implications of generative AI. This shift was reflected in the rise of negative sentiments and articles expressing *fear and concern* about AI, signaling a growing skepticism about the pace and direction of AI development.

While this study provides comparative insights on AI discourse in three Latin American countries, it also underscores the importance of regional perspectives in understanding global technological trends. While the three countries we analyzed generally followed broader global trends, we also identified interesting localized debates and regional trends that differentiate the discourse between countries. In Argentina, there was early engagement with the topic, with a strong focus on philosophical and ethical implications. In Colombia, the discourse often highlighted the intersection of AI with public safety, work automation, and its societal impact. Meanwhile, in Mexico, the focus was on cutting-edge technological advancements, practical applications of AI, and a persistent fascination with the technology. This interplay highlights the need for region-specific research to inform AI governance frameworks that are inclusive and adaptive to diverse realities.

Future research could expand this analysis by incorporating other media forms, such as social media or television, to capture a broader spectrum of public opinion. Additionally, exploring the perspectives of marginalized voices, including Indigenous communities and underrepresented groups, could provide a more comprehensive understanding of AI's societal impact. As AI continues to evolve, monitoring its portrayal in the media will remain critical for understanding and shaping its role in society.

References

Ahmadnia, S., Jordehi, A. Y., Hosseini Khasheh Heyran, M., Mirroshandel, S. A., & Rambow, O. (2024). Opinion Mining Using Pre-Trained Large Language Models: Identifying the Type, Polarity, Intensity, Expression, and Source of Private States. In *Proceedings of the 2024 Joint International Conference on Computational Linguistics, Language Resources and Evaluation (LREC-COLING 2024)* (pp. 12481–12495). ELRA and ICCL.

Baumgartner, F. R., & Jones, B. D. (2009). *Agendas and Instability in American Politics* (2nd ed.). University of Chicago Press.

Becerra, M., & Mastrini, G. (2017). *Periodismo y concentración de medios en América Latina*. Paidós.

Brown, T. B., Mann, B., Ryder, N., Subbiah, M., Kaplan, J., Dhariwal, P., et al. (2020). Language models are few-shot learners. In *Proceedings of the 34th International Conference on Neural Information Processing Systems (NIPS '20)* (pp. 1877–1901). Curran Associates Inc.

Canavilhas, J., Ioscote, F. C., & Gonçalves, A. V. (2024). Artificial Intelligence as an Opportunity for Journalism: Insights from the Brazilian and Portuguese Media. *Advances in the Social Sciences, 13*(11), 590–590. https://doi.org/10.3390/socsci13110590

Chuan, C.-H., Tsai, W., & Cho, S. (2019). Framing Artificial Intelligence in American Newspapers. In *Proceedings of the 2019 ACM Conference on Human Factors in Computing Systems* (pp. 339–344). https://doi.org/10.1145/3306618.3314285

Devlin, J., Chang, M.-W., Lee, K., & Toutanova, K. (2019). BERT: Pre-training of Deep Bidirectional Transformers for Language Understanding. In *Proceedings of the 2019 Conference of the North American Chapter of the Association for Computational Linguistics: Human Language Technologies* (pp. 4171–4186). Association for Computational Linguistics.

Fast, E., & Horvitz, E. (2016). Long-Term Trends in the Public Perception of Artificial Intelligence. *Proceedings of the AAAI Conference on Artificial Intelligence, 31*(1), 11295. https://doi.org/10.1609/aaai.v31i1.11295

Garvey, S., & Maskal, C. (2019). Sentiment Analysis of the News Media on Artificial Intelligence Does Not Support Claims of Negative Bias Against Artificial Intelligence. *OMICS: A Journal of Integrative Biology, 24*(6). https://doi.org/10.1089/omi.2019.0078

Hernandes, R., & Corsi, G. (2024). Artificial Intelligence in Brazilian News: A Mixed-Methods Analysis. *ArXiv Preprint.* https://doi.org/10.48550/arxiv.2410.17423

Hughes, S., & Márquez, M. (2017). Media, Politics, and Democracy in Mexico. In J. Iguíñiz & F. Darío (Eds.), *Media and Political Transitions in Latin America* (pp. 45–67). Oxford University Press.

Jacobs, R. N., & Townsley, E. (2011). *The Space of Opinion: Media Intellectuals and the Public Sphere.* Oxford University Press. https://doi.org/10.1093/acprof:oso/9780199736846.001.0001

Kelley, P., Yang, Y., Heldreth, C., Moessner, C., Sedley, A., Kramm, A., Newman, D., & Woodruff, A. (2021). Exciting, Useful, Worrying, Futuristic: Public Perception of Artificial Intelligence in 8 Countries. In *AIES '21: Proceedings of the 2021 AAAI/ACM Conference on AI, Ethics, and Society* (pp. 627–637). https://doi.org/10.1145/3461702.3462605

Kingdon, J. W. (1995). *Agendas, Alternatives, and Public Policies* (2nd ed.). HarperCollins.

Lawson, C. (2002). *Building the Fourth Estate: Democratization and the Rise of a Free Press in Mexico.* University of California Press.

Llorente, M., & Abad, F. (2020). Transformaciones en la prensa colombiana: El caso de Semana. *Revista Latinoamericana de Comunicación, 15*(2), 123–142.

McCombs, M. E., & Shaw, D. L. (1972). The Agenda-Setting Function of Mass Media. *Public Opinion Quarterly, 36*(2), 176–187. https://doi.org/10.1086/267990

Moriniello, F., Martí-Testón, A., Muñoz, A., Silva Jasaui, D., Gracia, L., & Solanes, J. E. (2024). Exploring the Relationship between the Coverage of AI in WIRED Magazine and Public Opinion Using Sentiment Analysis. *Applied Sciences, 14*(5), 1994. https://doi.org/10.3390/app14051994

Nader, K., Toprac, P., Scott, S., & Baker, S. (2022). Public Understanding of Artificial Intelligence Through Entertainment Media. *AI & Society, 39*, 713–726. https://doi.org/10.1007/s00146-022-01427-w. PMID: 35400854; PMCID: PMC8976224.

Newman, N., Fletcher, R., Eddy, K., Robertson, C. T., & Nielsen, R. K. (2023). *Reuters Institute Digital News Report 2023.* Reuters Institute for the Study of Journalism. https://doi.org/10.60625/risj-p6es-hb13

OpenAI. (2023). GPT-4 Technical Report. *ArXiv Preprint.* https://doi.org/10.48550/arXiv.2303.08774

OpenAI. (2025). *Models.* Retrieved January 4, 2025, from https://platform.openai.com/docs/models

Ouchchy, L., Coin, A., & Dubljević, V. (2020). AI in the Headlines: The Portrayal of the Ethical Issues of Artificial Intelligence in the Media. *AI & Society, 35*(4), 927–936. https://doi.org/10.1007/s00146-020-00965-5

Parratt, S. F., Chaparro, M. Á., & Martín, I. (2024). Spanish Media Coverage of Journalistic Artificial Intelligence: Relevance, Topics and Framing. *Revista Mediterránea de Comunicación: Mediterranean Journal of Communication, 15*(2), e25169. https://doi.org/10.14198/medcom.25169

Rey, G. (2012). *Periodismo y posconflicto: Claves de la prensa colombiana.* Debate.

UNESCO. (2022). *Recommendation on the Ethics of Artificial Intelligence.* United Nations Educational, Scientific and Cultural Organization.

Waisbord, S. (2013). *Reinventing Professionalism: Journalism and News in Global Perspective.* Polity.

Wankhade, M., Rao, A. C. S., & Kulkarni, C. (2022). A Survey on Sentiment Analysis Methods, Applications, and Challenges. *Artificial Intelligence Review, 55*(7), 5731–5780. https://doi.org/10.1007/s10462-022-10144-1

Zhang, W., Deng, Y., Liu, B., Pan, S., & Bing, L. (2024). Sentiment Analysis in the Era of Large Language Models: A Reality Check. In *Findings of the Association for Computational Linguistics: NAACL 2024* (pp. 3881–3906). Association for Computational Linguistics.

The Algorithmic Tyranny in the Gig Economy: National Strategies and Policy Implications in the Latin American Context

Alisa Petroff

Introduction

The last decade has seen a rapid development of the platform economy, and the COVID-19 pandemic has accelerated this trend (Joo & Shawl, 2021). While developed countries currently dominate the demand for gig workers, the demand in developing countries is growing at a much faster rate (Datta et al., 2023). One challenge when tackling this topic is related to the diversity of terminology which creates a certain ambiguity about the scope of the platforms and the implications for the actors involved (Petroff, 2022). Hunt and Samman (2019) distinguish between on demand platforms, managed digitally and performed offline, and crowdsourcing platforms, focused on specialized online assignments (from micro tasks to

A. Petroff (✉)
Universitat Oberta de Catalunya, Internet Interdisciplinary Institute (GenTic-IN3), Barcelona, Spain
e-mail: apetroff@uoc.edu

D. Ramírez Plascencia, R. M. Alonzo González (eds.), *Ethical and Legal Dilemmas of Artificial Intelligence in Latin America*,
https://doi.org/10.1007/978-3-031-86540-4_4

complex tasks) requiring a broad spectrum of skills and knowledge, often related to ICT sectors.

Regardless of its nature, platform economy is an innovative business model that creates economic value by using AI algorithms and digital technology (Zhou, 2022). However, this is a double-edged sword in terms of impact on labor relations. On one hand, matching algorithms allow platforms to connect workers to the tasks best suited to their skills and experience (Sutherland & Jarrahi, 2018). On the other hand, workers are constantly evaluated for their services and face a completely opaque system which lacks transparency about how people are assessed or how they obtain the job (Arruda & Pimenta, 2024). Therefore, a worker's rating, which is usually based on algorithms that combine the evaluations of customers, and the data collected from monitoring by the platform itself, involves considerable work-related consequences (ILO, 2020).

Furthermore, the relationship established between platforms and workers is asymmetrical and in the section of terms and conditions, workers are obliged to accept otherwise platforms deny any working relationship. This leads to a set of regulatory voids thanks to the "legal engineering," (Lehdonvirta, 2016) a process that creates legal grey areas where algorithmic decisions take the lead. As a result, labor platforms and employer companies gain twice as much from their business model. They win when they collect and process large amounts of user and workers' data, using algorithms to generate insights and to make decisions. At the same time, this datafication allows companies to price labor, creating predictions about how much they can pay, according to the prospects of economic dependence on workers (Sundararajan, 2017). In response to these tensions, workers display what Kellogg et al. (2019) call "algoactivism," that ranges from individual practical action to platform organizing, discursive framing, and legal mobilization. The resistance strategies under algorithm control include both individual and collective actions to navigate the existing constraints. An example of individual strategy relates to controlling the number of rejected orders, with a higher potential of failing or receiving bad feedback. Additionally, under the supervision of the rating and ranking system based on clients' feedback, client/worker relationships become vitally important to the worker's income (Sun, 2022).

In terms of collective actions, Sigg et al. (2024) propose a model aiming to increase the price of the services delivered through the platform. In their model, based on driving platforms like Uber, a pool of N drivers serves the requested deliveries, one per time-step. Drivers participating in the collective decline orders below a certain threshold pay. Non-participating drivers accept the request at any price. The platform returns any declined order to a random available worker at a slightly increased price, until a worker accepts the order. Moreover, the activities of trade unions focus on formulating ethical guidelines and principles about the impact of automation in the world of work. ITUC (2023) suggests that trade unions need to shift focus to a more practice-oriented approach on algorithmic transparency and accountability.

From this perspective, although activism is a necessary tool to empower workers in the gig economy, it is not enough to fight against the negative effects of the algorithms on working conditions. To address this issue, in spring 2024, the European Parliament and the Council launched the EU Platform Work Directive, an innovative political instrument which establishes a floor of individual and collective rights, including a number of advanced algorithmic management safeguards. For instance, Article 10(5) prohibits decisions to limit, suspend, or terminate the contractual relationship or the account, or "any other decision having equivalent effect," unless made by a human being. Moreover, Article 7(1)(f) prohibits processing of any personal data to derive information such as racial or ethnic origin, migration status, political opinions, religious or philosophical beliefs, disability, health, union membership, or sexual orientation, as well as the processing of any biometric data to establish a person's identity, as in the case of facial recognition in login practices. Article 12(3) introduces a ban on automated systems putting undue pressure on workers or creating risks to their safety and physical and mental health (Rainone & Aloisi, 2024).

Similarly, in 2019 the US government launched the Algorithmic Accountability Act (Congress.Gov, 2019). This bill requires specified commercial entities to conduct assessments of high-risk systems that involve personal information or make automated decisions, such as systems that use AI or machine learning. Specifically, high-risk automated decision systems include those that (1) may contribute to inaccuracy, bias, or discrimination; or (2) facilitate decision-making about sensitive aspects of consumers' lives by evaluating their behavior. Further, an automated-decision system, or information system involving personal data, is considered high-risk if it (1) raises security or privacy concerns, (2) involves the personal information of a significant number of people, or (3) systematically monitors a large, publicly accessible physical location.

Assessments of high-risk automated-decision systems must (1) describe the system in detail, (2) assess the relative costs and benefits of the system, (3) determine the risks to the privacy and security of personal information, and (4) explain the steps taken to minimize those risks, if discovered. Assessments of high-risk information systems involving personal information must evaluate the extent to which the system protects the privacy and security of such information.

Gig Economy in Latin America

At the turn of the twenty-first century, only 30 percent of Latin American citizens had access to the Internet (World Bank, 2017). In recent years, this number has doubled to 62 percent, but the gains are unevenly distributed (ITU, 2015). Despite the significant development, the region still lags behind Europe and North America in network speeds and full mobile access. Puerto Rico, Bahamas, St. Vincent & the Grenadines, Barbados, Argentina, Chile, and Uruguay have at least twice the internet penetration of impoverished nations such as El Salvador, Guatemala, Honduras, Nicaragua, and Haiti (Pick et al., 2020). Currently, the coverage gap is concentrated in remote areas or locations with a complex terrain, making network deployments economically challenging.

Regarding the digital economy, there is no representative data available on the quantitative importance of the platform jobs in Latin American countries, and the net-generation is even less clear when comparing new jobs against possible simultaneous job losses (for example, taxi drivers and delivery workers employed directly by restaurants, among others). However, estimations based on the partial information available point to job growth in both online web-based and location-based platforms in the region (CEPAL, 2021). While prior to the pandemic, the growth of digital platforms in Latin America was slow, the confinement measures that resulted from COVID-19 favored the increase in the use of the platforms to reach job opportunities (Azuara Herrera et al., 2020).

According to Bensusán and Santos (2021) platform work threatens to informalize the already highly informal labor markets in Latin America and to increase already high levels of inequality. The region's large technology corporations have found in this form of "entrepreneurship" an effective tool for the construction of a common sense functional to their interests and based on the idea of hyper individualized subjects, illusorily free of social conditioning, motivated by a diffuse idea of freedom, audacity, innovation, autonomy, risk and hyper-productivity (Ventrici, 2022).

Azuara et al. (2019)—In their study on Uber in Mexico, Chile, Brazil and Colombia—show that almost half the drivers had been informally employed before joining the company. However, income levels from digital platforms are a significant incentive, equivalent to three times the national minimum wage. Most of the digital workers are satisfied with their working conditions and only 40% of these workers stated that they would leave these jobs if they were offered other paid work.

In Argentina, only 55% of workers in the platform economy contribute to social security and for 61% of them it constitutes their main source of income (Madariaga et al., 2019). Just like in the case of Uber drivers, in general, most of the participants are workers at the beginning of their working trajectory with a high level of education (Bensusán & Santos, 2021). Despite counting on high levels of education, a recent study conducted in Argentina by ILO (2020) highlights that only 54% of the respondents affirm having seen the document with the terms and conditions of their service and 22% of these respondents affirm that they have not read them or think that the terms and conditions are not clear enough.

In Chile, workers earn more than the local minimum wage ($326,500 pesos per week with a 45-hour work schedule) before considering the costs. However, none of the platforms can reflect the pay of fair wage that includes gasoline expenses or the depreciation of their means of transportation (car, motorcycle, or bicycle) (Arriagada et al., 2021). In Uruguay, the study developed by Fairwork (2023) shows how the minimum wage in the delivery sector is $176.74 per hour, while the fair wage is about $229.77 per hour; and in the transportation sector, it is $182.59 per hour and $237.37 per hour respectively. In Ecuador, between 2019 and 2021, Glovo's algorithm modified its impact on workers' earnings by bundling two deliveries for a single compensation, reducing the payment for each delivery from 1.00 USD to 0.30 USD, and decreasing the value paid per kilometer travelled (Albornoz & Chávez, 2024).

In Colombia, many workers report a lack of communication channels with digital platforms, particularly in emergency situations. At the same time, the automation of the interactions with the platforms, based on natural language processing technologies, is making human-to-human interactions increasingly scarce. The platforms cover specific customers and parts of the city, but their complaints seem to have no effect on the platform. There are no replies or answers from the company in this medium.

Additionally, there is some distrust of the rating system amongst workers. This is perceived as surveillance by the platform. There is a perception that a bad review can be traced by the platform and be punished by blocking the account (Sanchez Vargas & Maldonado Castañeda, 2022).

Galperin and Greppi's (2017) study on the Nubelo platform, one of the largest online labor platforms serving Spanish-speaking countries, shows how the complex algorithmic process behind this platform reproduces structural inequalities and exposes migrants in Spain and Latin American countries to discriminatory practices embodied by the algorithms. This is particularly risky in the case of vulnerable groups, like Venezuelan migrants in Argentina, who represent 60% of the overall pull of workers in the gig economy in this country (Pereyra & Poblete, 2022).

Algorithms and Data Protection Regulations in Latin America

The developmental state literature can be seen as a benchmark for governmental efforts to move in the direction of greater state intervention in economic management with a vision to advancing social policy. This implies studying regulatory projects that challenge established policy approaches and entrenched interests. The technological innovations offered by platformization may cause just such a disruption and can therefore be studied as moments of potential for social change (Reilly, 2020).

In Latin America personal data regulations take a normative approach, which, much like the General Data Protection Regulation of the European Union, seeks to strike a balance between a personal need for privacy, and a corporate need to access personal data (Rodriguez & Alimonti, 2020). Despite these regulatory efforts, the algorithms of the digital platforms, essential for assigning a job and setting the price, operate as control mechanisms for these activities. Whoever takes that order does not always know how much they will be paid until the task is completed, which reflects an important information asymmetry (Fairwork, 2024). Additionally, political systems and practices in Latin America are remarkably accommodating for business interests, especially the narrow or individual interests of big business (Schneider, 2013, p. 148). In the following sections, I detail the regulations in different Latin American countries in an effort to map the advances in the legal frameworks in the main context of the region.

Argentina

In the case of Argentina, the entrance of digital platforms in the country relates to the financial deregulation promoted by Macri's government of 2015–2019 (Madariaga et al., 2019). Moreover, the development of the gig economy links with the dilution and weakening process of central institutions that promote ideas of entrepreneurship instead of reinforcing workers' rights to decent job opportunities (Haidar, 2020).

In 2018, the general strategic framework concerning the use of AI in Argentina was defined with the publication of its Digital Agenda for 2030. The agenda aims to take advantage of digital technologies for development and inclusion. It also established the construction of the National Artificial Intelligence Plan that looks at the ethical aspects of AI and the impact of technology in the world of work (ITUC, 2023). Although the plan was approved on December 9, 2019 (the last day of Macrista's presidency), it did not include an administrative structure or budgetary allocations for its development. At the end of 2019, with the arrival of Alberto Fernández to the presidency (2019–2023), the report was quickly forgotten (Vercelli, 2024).

Brazil

The general outlook of AI regulation in Brazil is marked by the Brazilian Artificial Intelligence Strategy and the national data protection law (Law N° 13,709/2018). While this strategy attempts to foster the use of AI by taking data protection and transparency into account, commercial secrets so far limit the transparency of algorithms in the private sector (ITUC, 2023). Currently, there are no legislative or regulatory projects on the gig economy, but the government is conducting a consultation that may result in substantial modifications in competition policy and in a new regulatory framework that impacts companies operating in the digital economy market (Volman et al., 2024).

Chile

Broader discussions on AI and algorithmic transparency are advancing quickly in this country thanks to the publication of a national strategy on AI by the Ministry of Science in 2021 (ITUC, 2023). More recently, on March 8, 2022, Chile became the first country in the Latin American region to launch a new legal framework to regulate new forms of work derived from the use of on-demand digital platforms. Chapter X adds "Work through Digital Labor Platforms" to the Special Contracts section of the Labor Code (ISSA, 2023). More specifically, the law n° 21.431 "aims to regulate the situation of digital platforms in Chile," defined as "any digital infrastructure whose purpose is to create, organize and control, through algorithms and people, an interaction space where individuals can exchange information, goods or services." It establishes a service provision contract by controlling the terms and conditions: prices and rates: contact and coordination criteria between worker and user: protection of personal data: and maximum connection times. In consequence, it reduces the variations derived from the algorithms developed by the platforms (Inter-American Development Bank, 2022).

More particularly, Article 152d E. includes the prohibition of discrimination by automated decision-making mechanisms. In the implementation of the algorithms, the digital company must respect the principle of equality and non-discrimination. To this end, it will take all necessary measures and safeguards to avoid any type of discrimination among workers, particularly in the assignment of work, offer of bonuses and incentives, and calculation of remuneration among others. Additionally, the competent authority is allowed to supervise and have access to the programming of the algorithm, to receive complete and sufficient explanations of its operation and the data with which it was trained (González Candia et al., 2024).

The Chilean platform law could have done more to strengthen workers' rights in areas such as algorithmic management and workers' access to algorithm criteria (Arriagada et al., 2023). The broad set of factors considered in the formation of prices, and the variability inherent in algorithmic price management makes it difficult for workers to track that payments are correct, even when applications have a record of payments for service (FLACSO, 2024). By following the example of the Spanish "rider law," which grants platform worker organizations the right to demand access to algorithm information, the Chilean law could have better protected workers (Leyton et al., 2022).

Colombia

At the beginning of 2023, the Ministry of Labor in Colombia presented an extensive Labor Code reform to congress that would have regulated delivery work through online platforms. Intended changes included granting workers all rights and benefits currently in place for dependent workers while under a special contract, algorithm transparency in automated decision-making, and the development of a registry of all platform delivery workers. The reform was not approved due to insufficient quorum for debate (eight out of 21 commission members), preventing its discussion before the end of the legislative session. However, the ministry is expected to present a new project for the upcoming session (ISSA, 2023).

Ecuador

In January 2023, a bill-regularizing platform was approved for a second debate by the National Assembly of Ecuador's Commission of Workers' Rights and Social Security. Notably, it would change article 1 of the Labor Code by incorporating transparency provisions requiring employers to inform workers of the parameters on which algorithms or artificial intelligence systems affecting decision making (working conditions, access to and maintenance of employment and creation of profiles) are based (ISSA, 2023).

Mexico

President Claudia Sheinbaum, on October 16, 2024, signed the initiative to reform the Federal Labor Law (LFT) with the aim of guaranteeing labor rights to digital platform workers. Consequently, the reform initiative contemplates additions to the LFT, which includes new sections to articles 49, 50, and 127, as well as the incorporation of a new chapter on work on digital platforms. Among the main topics, the reform tackles the algorithmic management policy. More specifically, companies will need to have clear policies on the use of algorithms that affect task allocation and provide information on how these policies influence the conditions of users (Holland and Knight, 2024).

From the perspective of workers, the issue of transparency is vital. Omar Nacib Esteban Fuentes, general director of Social Welfare of the Ministry of Labor and Social Welfare (STPS), stressed that the Mexican government has been inspired by initiatives such as the European Union directive for work on digital platforms, which promotes "algorithmic management" with the intention of offering workers a basic understanding of how algorithms operate. The STPS proposes a document that explains in simple terms the rules under which tasks are assigned, hours are counted, and evaluations are distributed (Riquelme, 2024).

At the Impact of Platforms forum in the framework of digital entrepreneurship in Mexico, organized by the Law and Artificial Intelligence Research Line (LIDIA) of the UNAM, in collaboration with the Latin American Internet Association, Pablo Pruneda, coordinator of LIDIA, was clear in warning that the objective of achieving greater transparency on the platforms has a good intention, but that algorithmic transparency policies can fall into the error of trying to regulate an aspect of the business model that is essential for its operation (Riquelme, 2024).

Finally, the Investigative Authority (IA) of the Mexican Federal Economic Competition Commission (COFECE) preliminarily determined that there are no conditions of effective competition in the retail e-commerce market in Mexico. To eliminate the barriers related to the lack of the transparency in the operation of algorithms, a set of corrective measures have been proposed: (1) creating a section in the portal that informs sellers of the criteria of management algorithms to provide transparency and certainty and (2) modifying the criteria of the management algorithms based on preliminary discussions and observations between the economic agents involved (Volman et al., 2024).

Uruguay

More broadly, there are no formal regulations on AI and algorithmic transparency in Uruguay. However, there is an incipient discussion on algorithmic transparency, mainly promoted by governmental entities, specifically by the Agency for the Development of e-Government and the Information and Knowledge Society (AGESIC) and its AI strategy for digital government (ITUC, 2023).

From this perspective, this country is currently in the process of developing its AI National Strategy. Although specific laws have not yet been crafted, Article 16 of Uruguay's Data Protection Law, gives individuals the right to avoid decisions made solely by automated systems if those decisions have significant effects on them, such as in credit scoring or job applications (International Trade Administration, 2024). Additionally, in October 2022 the Ministry of Labor and Social Security presented to the Parliament a law project aiming to regulate work through digital platforms. One of the requirements included the obligation of digital platforms to respect the principle of equality and non-discrimination in the implementation of algorithms. Moreover, companies must inform workers about the existence of automated monitoring systems, as well as automated decision-making systems. The proposal also established the right to an explanation in favor of workers, in relation to automated decisions taken by the company, and individuals must be designated physical contact in order to clarify such decisions. However, since October 2022, the initiative has not been further developed and put into practice (Fairwork, 2023).

It calls for transparency in algorithm and monitoring systems, focusing on automated decision-making features, grants workers the right to contest significant changes in working conditions, and requires platforms to implement preventive measures based on assessed risks (ISSA, 2023).

Conclusions

We have witnessed how "the future of work" has gone from a thought or a provocative phrase to a political statement and a real matter of concern for politicians, unions, experts and the private sector. These actors have been heavily influenced by the future promised by digital platforms and algorithmic management: flexibility, control, low costs, and rising profits. These imaginaries have been reproduced by local media, particularly by business press, who have kept the interest in such matters alive through opinion and news and by staging different public discussions about the future of work and the professions that will be required in a Fourth Industrial Revolution scenario (Sanchez Vargas & Maldonado Castañeda, 2022).

Algorithmic management and work organization through technological platforms are expanding to different markets and public services. Five are the essential characteristics of this type of management: (1) the continuous monitoring of the behavior of workers; (2) the constant evaluation of workers through assessments by customers; (3) decisions are apparently made and implemented without human intervention; (4) worker interaction through the application, thereby hindering the possibility of workers being able to communicate with each other and to personally appeal certain decisions; and (5) a certain lack of transparency about how the algorithm works, due to both competitive commercial practices and the very adaptability of the algorithm (ILO, 2020; Berg et al., 2018).

In view of these features, new changes will radically transform labor dynamics and the way social rights are organized and guaranteed (Fairwork, 2024). Additionally, the humanitarian crisis triggered by COVID-19 compelled policymakers to modify labor regulations to mitigate widespread layoffs in Latin American countries (Albornoz & Chávez, 2024). Since the beginning of the twenty-first century, Latin American countries have embarked on ambitious digital transformation processes. The most recent Digital Agenda for Latin America (eLAC2024), adopted at the Eighth Ministerial Conference on the Information Society in Montevideo 2024, establishes a regional policy framework for digital transformation for 2023 and 2024 (UN, 2024). However, this chapter shows how different counties of the region have unevenly adopted legislations to regulate algorithms in the digital platforms.

Based on these assumptions, some recommendations should be considered.

- Bensusán and Santos (2021) highlights the need to establish legal responsibility for companies using algorithms to organize work. Different organizations, entities, and collectives engage in activities on algorithmic transparency and accountability in the world of work. Traditional labor unions and union confederations are by far the most important actors engaging in empowerment, capacity-building, campaigning, advocacy, and formulating regulatory demands (ITUC, 2023).
- Muller (2020) proposes common law causes of action sounding in contract and tort jurisprudence as a potential response to algorithmic harms to platform workers. The law is slow to adapt to structural economic change. The modern worker protection regimes were not drafted overnight but rather came in the wake of decades of litigation between workers and employers, with courts laying down the principles that would animate these laws through case-by-case adjudication. In the present moment, where the reach and import of machine learning algorithms continues to expand dramatically while regulations and laws pertaining to them remain scant, the courts might play a similar role in extending the law from common principles.
- The terms and conditions should clearly establish the criteria for assigning and performing the service and explain not only how the algorithms that establish the assignment of tasks operate but also the following: how the remuneration is calculated, the form of payment, how the workers' data is collected and used, the mode and impact of the score that is given, the procedure in the event that a product is returned or rejected by the final consumer, the causes for blocking or

disconnection and any other essential condition for performing the activity (ILO, 2020).

- I argue that legislation on the topic of algorithms in the digital economy should respond to context-based features, considering also the current advances made by different governments in the region, the extent to which the gig economy has been developed in recent decades and the set of opportunities and challenges already raised by algorithms in these forms of work.

References

Albornoz, M. B., & Chávez, H. (2024). The Challenges of Gig Economy and Fairwork in Ecuador. *Digital Geography and Society, 6*, 100073. https://doi.org/10.1016/j.diggeo.2023.100073

Arriagada, A., Bonhomme, M., Ibáñez, F., Leyton, J., Ustek-Spilda, F., Graham, M., & Bertolini, A. (2021). Fairwork Chile Puntuaciones 2021: Nuevo informe sobre la economía de plataformas en América Latina.

Arriagada, M., Bonhomme, F., & Ibáñez Leyton, J. (2023). The Gig Economy in Chile: Examining Labor Conditions and the Nature of Gig Work in a Global South Country. *Digital Geography and Society, 5*. https://doi.org/10.1016/j.diggeo.2023.100063

Arruda, E. P., & Pimenta, D. (2024). Challenges and Implications of Microwork in the Age of Artificial Intelligence: A Global Socioeconomic Analysis. *Human Resources Management and Services, 6*(2), 3452. https://doi.org/10.18282/hrms.v6i2.3452

Azuara Herrera, O., Fazio, M. V., Hand, A., Keller, L., Rodríguez Tapia, C., & Silva Porto, M. T. (2020). *El futuro del trabajo en América Latina y el Caribe: ¿Cómo puede la tecnología facilitar la recuperación del empleo tras el COVID-19?* Banco Interamericano de Desarrollo. Retrieved from https://publications.iadb.org/es/el-futuro-del-trabajo-en-america-latina-y-el-caribecomo-puede-la-tecnologia-facilitar-la

Azuara, O., González, S., & Keller, L. (2019). *¿Quiénes son los conductores que utiliza las plataformas de transporte en América Latina? Perfil de los conductores de Uber en Brasil, Chile, Colombia y México.* División de Mercados Laborales del Banco Interamericano de Desarrollo.

Bensusán, G., & Santos, H. (2021). Chapter 10: Digital Platform Work in Latin America: Challenges and Perspectives for its Regulation. In *En Work and Labour Relations in Global Platform Capitalism.* Edward Elgar Publishing. https://doi.org/10.4337/9781802205138.00020

Berg, J., Furrer, M., Harmon, E., Rani, U., & Silberman, S. (2018). *Digital Labour Platforms and the Future of Work: Towards Decent Work in the Online World*. International Labour Organization.

CEPAL. (2021). *Decent Work for Platform Workers in Latin America*. Retrieved from https://repositorio.cepal.org/server/api/core/bitstreams/5484c1bf-c259-4a9a-aa98-22db13d52a8b/content

Congress.gov. (2019). *Algorithmic Accountability Act of 2019*. Retrieved from https://www.congress.gov/bill/116th-congress/house-bill/2231/all-info#:~:text=%2F10%2F2019)-,Algorithmic%20Accountability%20Act%20of%202019,artificial%20intelligence%20or%20machine%20learning

Datta, N., Rong, C., Singh, S., Stinshoff, C., Iacob, N., Nigatu, N. S., Nxumalo, M., & Klimaviciute, L. (2023). *Working Without Borders: The Promise and Peril of Online Gig Work*. World Bank.

Fairwork. (2023). *Fairwork Uruguay. Puntuaciones 2023. La precariedad laboral versus la promesa de flexibilidad y autonomía de la economía de plataformas*. Retrieved from https://fair.work/wp-content/uploads/sites/17/2024/02/Fairwork-Uruguay-Ratings-2023-ES.pdf

Fairwork. (2024). *Vivir como trabajador de plataformas en Chile: Informe 2024*. Retrieved from https://www.ciperchile.cl/2024/09/16/vivir-como-trabajador-de-plataformas-en-chile-informe-2024/

FLACSO. (2024). *Estudio de descripción de las condiciones de trabajo y empleo en el trabajo de plataformas digitales, conforme a lo dispuesto por la Ley n°21.431*. Retrieved from https://www.subtrab.gob.cl/wp-content/uploads/2024/05/7-Resumen-Ejecutivo.-Trabajo-en-Plataformas-Digitales-de-Servicios-Ley-21.431.pdf

Galperin, H., & Greppi, C. (2017). Geographical Discrimination in Digital Labor Platforms. In M. Graham (Ed.), *Digital Economies at Global Margins* (pp. 295–318). MIT Press.

González Candia, J., Hernández Román, G., Mardones, E. M. R., & Von Der Meden, O. Z. (2024). Regulación laboral para las plataformas digitales en Chile: Descripción de la Ley N° 21.431. *Gaceta Laboral, 28*(2), 74–94.

Haidar, J. (2020). *La configuración del proceso de trabajo en las plataformas de reparto en la Ciudad de Buenos Aires. Un abordaje multidimensional y multi-método*. Instituto de Investigaciones Gino Germani (IIGG), Facultad de Ciencias Sociales, Universidad de Buenos Aires (FSOC—UBA). Informes de Coyuntura N°11, Octubre 2020.

Holland and Knight. (2024). *Iniciativa de reforma laboral en México para regular a trabajadores de plataformas digitales*. Retrieved from https://www.hklaw.com/es/insights/publications/2024/10/iniciativa-de-reforma-laboral-en-mexico-para-regular-a-trabajadores

Hunt, A., & Samman, E. (2019). *Gender and the Gig Economy: Critical Steps for Evidence-Based Policy*. Working Paper 546. Overseas Development Institute.

Inter-American Development Bank. (2022). *La regulación del trabajo en plataformas en Chile: ¿Nuevo paradigma para la región?* Retrieved from https://blogs.iadb.org/trabajo/es/la-regulacion-del-trabajo-en-plataformas-en-chile-nuevo-paradigma-para-la-region/

International Labour Organization (ILO). (2020). *Work on Delivery Platforms in Argentina: Analysis and Policy Recommendations.* Retrieved from https://www.ilo.org/sites/default/files/wcmsp5/groups/public/@americas/@ro-lima/@ilo-buenos_aires/documents/publication/wcms_765150.pdf

International Social Security Association (ISSA). (2023). *Platform Workers and Social Security: Recent Developments in the Americas.* Retrieved from https://www.issa.int/analysis/platform-workers-and-social-security-recent-developments-americas

International Telecommunication Union (ITU). (2015). *ICT Facts and Figures: The World in 2015.* ITU.

International Trade Administration. (2024). *Digital Economy.* Retrieved from https://www.trade.gov/country-commercial-guides/uruguay-digital-economy#:~:text=Digital%20Services%20Law%20(Law%20N,to%20improve%20accessibility%20and%20efficiency

International Trade Union Confederation (ITUC). (2023). *Algorithmic Transparency and Accountability in the World of Work.* Retrieved from https://algorithmwatch.org/en/wp-content/uploads/2023/02/2023_AlgorithmWatch_ITUC_Report.pdf

Joo, B. A., & Shawl, S. (2021). COVID-19 Pandemic and the Rising Gig Economy: An Emerging Perspective. *Global Economics Science, 2*(3), 16–23. https://doi.org/10.37256/ges.232021917

Kellogg, K., Valentine, M., & Christin, A. (2019). Algorithms at Work: The New Contested Terrain of Control. *Academy of Management Annals, 14*(1), 366–410. https://doi.org/10.5465/annals.2018.0174

Lehdonvirta, V. (2016). Algorithms that Divide and Unite: Delocalisation, Identity and Collective Action in "Microwork". In *Space, Place and Global Digital Work* (pp. 53–80). https://doi.org/10.1057/978-1-137-48087-3_4

Leyton, J., Arriagada, A., Bonhomme, M., & Ibáñez, F. (2022). Fairwork Policy Brief: New Regulation of Platform Work in Chile: A Missed Opportunity? *The Fairwork Project.* Available at: https://fair.work/wp-content/uploads/sites/17/2022/03/Policy-brief.-New-regulation-of-platform-work-in-Chile-a-missed-opportunity-EN.pdf

Madariaga, J., Buenadicha, C., Molina, E., & Ernst, C. (2019). *Economía de plataformas y empleo. ¿Cómo es trabajar en una APP en Argentina?* CIPPEC, BID, OIT.

Muller, Z. (2020). Algorithmic Harms to Workers in the Platform Economy: The Case of Uber. *Columbia Journal of Law and Social Problems, 53*(2).

Pereyra, J., & Poblete, L. (2022). Desafíos de la regulación del trabajo de plataformas digitales de reparto en Argentina. *Revista Temas Sociológicos, 31*, 359–391. https://doi.org/10.29344/07196458.31.3016

Petroff, A. (2022). Crowdsourcing in ICT Fields: The Link Between Mobility and Digital Labor Markets in Highly Skilled Tasks. In *The Palgrave Handbook of Global Social Change.*

Pick, J., Sarkar, A., & Parrish, E. (2020). The Latin American and Caribbean Digital Divide: A Geospatial and Multivariate Analysis. *Information Technology for Development, 27*(2), 235–262. https://doi.org/10.1080/02681102.2020.1805398

Rainone, S., & Aloisi, A. (2024). The EU Platform Work Directive. In *ETUI*. The European Trade Union Institute. Retrieved from https://www.etui.org/publications/eu-platform-work-directive

Reilly, K. (2020). Platform developmentalism: Leveraging Platform Innovation for National Development in Latin America. *Internet Policy Review, 9*(4), 1–29. https://doi.org/10.14763/2020.4.1538

Riquelme, R. (2024). ¿Regulación a plataformas digitales viola la secrecía de los algoritmos? *El Economista*. Retrieved from https://www.eleconomista.com.mx/capital-humano/regulacion-plataformas-digitales-viola-secrecia-algoritmos-20241113-734005.html

Rodriguez, K., & Alimonti, V. (2020). *A Look-Back and Ahead on Data Protection in Latin America and Spain.* [Blog post]. Electronic Frontier Foundation Deeplinks. Retrieved from https://www.eff.org/deeplinks/2020/09/look-back-and-ahead-data-protection-latin-america-and-spain

Sanchez Vargas, D. Y., & Maldonado Castañeda, O. (2022). 'My Boss, the App': Algorithmic Management and Labour Process in Delivery Platforms in Colombia. In *Future of Work(ers)*. SCIS Working Paper Number 36. Southern Centre for Inequality Studies, Wits University. https://doi.org/10.54223/uniwitwatersrand-10539-33459

Schneider, B. R. (2013). *Hierarchical Capitalism in Latin America: Business, Labor, and the Challenges of Equitable Development.* Cambridge University Press. https://doi.org/10.1017/CBO9781107300446

Sigg, D., Hardt, M., & Mendler-Dünner, C. (2024). Decline Now: A Combinatorial Model for Algorithmic Collective Action. https://doi.org/10.48550/arXiv.2410.12633

Sun, Y. (2022). The Algorithm-Based Platform Economy: A New Terrain of Control and Resistance. In *Proceedings of the 2022 International Conference on Science and Technology Ethics and Human Future (STEHF 2022).*

Sundararajan, A. (2017). *The Sharing Economy: The End of Employment and the Rise of Crowd-Based Capitalism.* MIT Press.

Sutherland, W., & Jarrahi, M. H. (2018). The Sharing Economy and Digital Platforms: A Review and Research Agenda. *International Journal of Information Management, 43*, 328–341. https://doi.org/10.1016/j.ijinfomgt.2018.07.004

United Nations (UN). (2024). *E-Government Survey 2024: Accelerating Digital Transformation for Sustainable Development*. Retrieved from https://desapublications.un.org/sites/default/files/publications/2024-09/%28Web%20version%29%20E-Government%20Survey%202024%201392024.pdf

Ventrici, P. (2022). Capitalismo de plataformas en América Latina. *Asociación Internacional de Sociología, Diálogo Global, 12*(2), 8–14.

Vercelli, A. (2024). Regulaciones e inteligencias artificiales en Argentina. *Inmediaciones de la Comunicación, 19*(1), 52–74. https://doi.org/10.18861/ic.2024.19.1.3549

Volman, V., Navarro, D., & Frene, L. (2024). *New Regulations for Marketplaces and Other Digital Platforms: Latest Updates in Argentina, Brazil, and Mexico*. International Bar Association. Retrieved from https://www.ibanet.org/new-regulations-for-marketplaces-and-other-digital-platforms-argentina-brazil-mexico

World Bank. (2017). *Uruguay's Digital Development Policy*. Retrieved from https://www.effectivecooperation.org/system/files/202106/GDI%20Delivery%20Note%20Uruguay%20Digital%20Development.pdf

Zhou, Y. (2022). An Overview of Platform Economy and Its Algorithms: Strengths, Weaknesses and Regulatory Initiatives. *BCP Business & Management, 19*, 158–161. https://doi.org/10.54691/bcpbm.v19i.738

Mapping the Future of AI Regulation in Latin America: Civil Society Perspectives on Brazil's Pilot AI Regulatory Sandbox

Kenzo Soares Seto

Introduction

This chapter aims to analyze the proposals made by Brazilian society during the public consultation organized by the Brazilian Data Protection Authority (Autoridade Nacional de Proteção de Dados, ANPD) on the creation of a regulatory sandbox for artificial intelligence (AI). Drawing on contributions from diverse actors, it is possible to outline the main points of consensus and controversy regarding the risks and opportunities surrounding the development and adoption of AI in one of the leading countries of the Global South and the largest economy in Latin America.

As the development and adoption of AI systems accelerates worldwide, debates about the legal regulation of these systems and the construction of governance mechanisms have become increasingly central, as has the role of civil society in the design and implementation of these institutional

K. Soares Seto (✉)
Federal University of Rio de Janeiro (UFRJ), Rio de Janeiro, Brazil
e-mail: kenzo.seto@eco.ufrj.br

D. Ramírez Plascencia, R. M. Alonzo González (eds.), *Ethical and Legal Dilemmas of Artificial Intelligence in Latin America*,
https://doi.org/10.1007/978-3-031-86540-4_5

frameworks. Civil society engagement, together with academic research, has primarily focused on legislative debates on AI regulation, with particular attention to pioneering initiatives in Europe (Crum & Coglianese, 2024; Gstrein et al., 2024; Hartmann et al., 2024) and, more recently, in the Global South (Belli, 2021; Wall et al., 2021; Derechos Digitales et al., 2024; Índice Latino-americano de Inteligência Artificial, ILIA, 2024). In this regard, numerous authors highlight the risks that AI poses, especially in the Global South, such as increasing algorithmic discrimination, data exploitation, digital colonialism, and state surveillance in the absence of robust regulation (Chinmayi, 2020; Wall et al., 2021).

In this context, efforts are underway in Latin America to balance the risks and opportunities of AI through public governance: 38 national legislative initiatives are under discussion in different countries, although few have established fully consolidated regulatory frameworks (ILIA, 2024). However, one of the main limitations of civil society engagement when limited to legislative processes is that parliamentary debates often emphasize broad values such as transparency, accountability and ethical principles, but often lack empirical evidence on how to translate these principles into precise methods for assessing the level of risk posed by specific AI applications (Guha et al., 2024).

As a result, legislation may create risk assessment systems that prove difficult to implement and provide vague assurances of rights without practical mechanisms for enforcement (Derechos Digitales et al., 2024). For example, decisions on whether to classify certain AI systems as high or medium risk for society, and thus determine the scope of regulatory requirements, may depend less on technical criteria than on the lobbying power of certain companies. This concern has been echoed, for example, by segments of civil society in Brazil, who have criticized the exclusion of social media recommendation and moderation systems from the high-risk category in the country's proposed AI regulatory framework (Lira, 2024).

In contrast, the global trend of establishing regulatory sandboxes for AI (Organisation for Economic Co-operation and Development, OECD, 2023) offers unique advantages over purely legislative debates, but has received comparatively less scholarly attention in Latin America than legislative discussions (ANPD, 2023a). AI regulatory sandboxes are controlled environments established by regulators to allow for the testing and prototyping of novel AI systems under temporary relaxations of certain rules, allowing for a calculated tolerance of risk (OECD, 2023). These sandboxes allow for an experimental, hands-on approach to regulation, where developers can test and evaluate the effectiveness of AI applications prior to large-scale deployment and under the close supervision of regulatory authorities (ANPD, 2023a). Therefore, regulatory sandboxes allow policymakers, developers, and civil society to observe technical and empirical evidence on the risks and benefits of AI systems. They also provide insights into the impact of public regulation on technological development, making it easier to assess whether existing legal and governance frameworks are sufficient or need further refinement.

Nevertheless, regulatory sandboxes involve a temporary, localized relaxation of rules designed to protect society from the potentially harmful effects of AI. As such, their design and management require careful consideration, and as they are debated by civil society, they offer a privileged vantage point for examining fears, aspirations, and controversies related not only to the broader implications of AI, but also to the specific challenges of local AI development, particularly in countries of the Global South.

In this regard, a significant challenge is the development of robust, clear, and auditable mechanisms for citizen participation in the formulation and validation of national AI strategies, sandboxes, and regulations—processes that remain largely absent in most countries in Latin America (ILIA, 2024). Numerous voices within Latin American civil society, such as the AISur coalition, which comprises dozens of civil society organizations across the continent, have raised concerns about limited opportunities for public participation in these regulatory processes, as well as about lobbying efforts by the tech industry to downplay the harmful effects of technologies such as facial recognition and the unauthorized use of biometric data (Derechos Digitales et al., 2024). At the same time, other actors in Latin America regard investment in AI as an avenue for overcoming economic underdevelopment (Ovanessoff & Plastino, 2017 and worry that excessive regulation could hinder the emergence of local enterprises and technological innovation (Wall et al., 2021).

Within this landscape of competing interests, it becomes crucial to examine how the ongoing regulatory debates materialize in a specific national context. Brazil offers an illuminating case: as one of the leading AI hubs in the Global South, the country's regulatory discussions and civil society engagement reveal both the opportunities and the challenges in shaping robust AI governance. The following sections therefore focus on Brazil's pivotal role in the Latin American AI landscape, analyzing its existing regulatory environment, legislative proposals, and tradition of multi-stakeholder participation.

Brazil's Regulatory Landscape for AI

Brazil, the largest economy in Latin America, leads the region in AI investment, research, and development (ILIA, 2024; Maslej et al., 2023) and ranks 38th globally in the Global AI Index (Maslej et al., 2023). It holds the 12th position worldwide in AI-related scientific research, the highest ranking among Latin American countries (Machado & Winter, 2023). However, in the Latin American Artificial Intelligence Index, Brazil falls slightly behind Chile, which ranks first, due to internal disparities in key AI enabling factors such as digital literacy, internet access, and regulatory gaps (ILIA, 2024).

A central issue is that Brazil still operates in an environment of legal and regulatory uncertainty around AI. In 2021, the government published the Brazilian Artificial Intelligence Strategy (Estratégia Brasileira de Inteligência Artificial, EBIA), which outlines principles for local AI development and adoption, such as transparency, accountability, and inclusivity. However, it lacks legal enforceability, and critical issues such as algorithmic bias, accountability for AI-driven decisions, and data sovereignty remain inadequately addressed (Zanatta & Rielli, 2024).

Additionally, the main Brazilian legislative proposal to establish comprehensive rules for the development and use of AI, Bill No. 2338/2023, has yet to be approved by the National Congress, although discussions have been ongoing since 2021. These discussions highlight the complexity of building a regulatory framework that is appropriate to the Brazilian reality, although inspired by international frameworks such as the European Union's (EU) AI Law. Unlike the EU's primarily risk-based model, Brazil emphasizes a rights-based framework alongside risk considerations, with an emphasis on guaranteeing unprecedented rights (Mendes & Kira, 2023).

In this context, Brazil serves as a notable example within the Global South, aiming to transcend the so-called "Brussels effect," where European legal norms are often uncritically adopted elsewhere (Seto, 2021), thus challenging the institutional reproduction of North-South political and epistemological hierarchies. If approved, the Brazilian bill could become the first to require mandatory explainability and output transparency, especially for AI systems classified as high risk. Under these rules, developers would need to clarify how and why certain decisions were made, thus mitigating potential biases (Lira, 2024; Zanatta & Rielli, 2024). Other Brazilian provisions that go beyond the EU's AI Act include energy efficiency and environmental sustainability of AI, as well as specific provisions for generative AI, such as the requirement to disclose when content is generated by AI (Lira, 2024).

The Bill No. 2338/2023 also pioneeringly addresses the issue of copyright in training data, requiring artists and authors to be notified when their works are used to train AI, granting them veto rights and fair compensation (Zanatta & Rielli, 2024). In addition, instead of centralizing governance in a single authority, as in the case of the European AI Office, Brazil proposes a National System for the Regulation and Governance of AI (Sistema Nacional de Regulação e Governança de Inteligência Artificial, SIA). This would create a collaborative framework that would include Brazilian sectoral regulators, such as those in telecommunications, health, and finance, to regulate AI applications in those sectors, with the ANPD as the overarching coordinating body. Oversight would also be provided by a multi-stakeholder council, the Permanent Council for Artificial Intelligence Regulatory Co-operation (Conselho Permanente de Cooperação Regulatória em Inteligência Artificial, CRIA), and informed by a technical committee of AI experts, the Technical Committee of AI Experts (Comitê Técnico de Especialistas em Inteligência Artificial, CECIA) (Zanatta & Rielli, 2024).

Multi-stakeholder Governance: The Controversial Voices of Society

Brazil's approach to AI regulation is grounded in a longstanding tradition of multi-stakeholder governance in the digital sphere. This tradition began in 1995 with the creation of the Brazilian Internet Steering Committee (Comitê Gestor da Internet no Brasil, CGI.br), a pioneering body that brought government, academia, the private sector, and civil society together into internet policy deliberations. It was further strengthened by the 2014 Brazilian Civil Rights Framework for the Internet (Marco Civil da Internet), the country's "Internet Constitution" (Conceição & Perrone, 2022), which emerged from extensive public consultation and enshrined a rights-based framework for internet governance. In 2018, the General Data Protection Law (Lei Geral de Proteção de Dados, LGPD) introduced data protection standards aligned with Brazil's context and created the ANPD to oversee their implementation, alongside another multi-stakeholder council, the National Data Protection Council (Conselho Nacional de Proteção de Dados, CNPD).

Nevertheless, the discussion of Bill No. 2338/2023 is an example of how the participation of different sectors of society signifies the constant confrontation of diverse viewpoints. For instance, the bill's proposed copyright provisions for generative AI resulted from concerted efforts by hundreds of artists, musicians, and authors, yet the Brazilian Software Companies Association (Associação Brasileira das Empresas de Software, ABES) regards the mandatory notification of creators as a barrier to Brazil's competitiveness in AI (Lira, 2024). While some see the bill's enhanced individual rights and civil liability provisions covering both AI providers and operators as a major step forward, others argue that focusing solely on risk will stifle innovation. This is the position of some influential voices in Brazilian civil society, such as Ronaldo Lemos, founder of the Institute for Technology and Society (Instituto de Tecnologia e Sociedade, ITS) and a key figure behind Brazil's Marco Civil da Internet (Lira, 2024).

Such critiques align with broader concerns in the literature about the unintended consequences of over-regulation. Guha et al. (2024) highlight how overly stringent rules can stifle innovation, raise compliance costs for smaller players—such as local startups—and therefore inadvertently solidify dominant market positions. In addition, policy proposals sometimes mandate measures that are not technically feasible. For example, requiring developers to provide explainability or reduce training datasets for intellectual property reasons may undermine model performance or accuracy, ultimately to the detriment of users (Guha et al., 2024).

In response to these challenges, Bill No. 2338/2023 envisages regulatory sandboxes as a means to simultaneously assess the risks posed by both AI models and their regulation (Zanatta & Rielli, 2024). However, this initiative is also mobilizing controversy in Brazilian civil society: critics argue that such sandboxes add another layer of bureaucracy and costs, especially for emerging companies (Lira, 2024), while proponents see them as an essential mechanism to reduce the information asymmetry between regulators and AI developers. They provide developers with a direct channel to the regulator, generate insights for regulation, and foster an environment of trust and legal certainty based on collaborative learning.

Moreover, supporters of the sandbox consider it a fundamental methodology for addressing one of the key challenges of regulation: defining the risks associated with each system to uphold the primary principle of Bill No. 2338/2023: "to ensure safe and reliable systems that serve humanity, democratic governance, and scientific and technological progress" (2023, p. 1). In this broader context, it is important to recognize that the concept of a regulatory sandbox for AI is not unique to Brazil's proposal but rather reflects both the country's multi-stakeholder tradition and its entry into a global trend in AI regulation.

The Rise of AI Regulatory Sandboxes

The concept of sandboxes originates from software development practices, where it refers to a controlled and isolated environment used to safely test code without impacting the broader system, thereby minimizing the risk of unintended consequences (ANPD, 2023a). Based on this principle, regulatory sandboxes were first applied in the financial technology sector as controlled environments that allow firms to test innovative services under regulatory oversight. These sandboxes have three main features: (1) they define testing zones for a limited period of time; (2) they allow for regulatory flexibility by offering firms protection from immediate penalties for non-compliance with certain rules; and (3) they aim to generate practical insights and evidence to inform future legislation (Bundesministerium für Wirtschaft und Energie, BWE, 2019). This approach gained global prominence with the launch of the United Kingdom (UK) Financial Conduct Authority's (FCA) sandbox program in 2015. By 2023, more than 57 countries had implemented regulatory sandboxes to balance the need for innovation with the demands of effective regulation (OECD, 2019).

In the context of AI, regulatory sandboxes have been adopted by dozens of national data protection authorities (DPAs), agencies similar to Brazil's ANPD, over the past decade. Singapore's Personal Data Protection Commission (PDPC) was an early pioneer, incorporating AI testing into its sandbox framework starting in 2017. Similar initiatives were subsequently launched in countries such as the United Kingdom, Norway, France, Japan, and Spain, generally focusing on concerns related to data protection, ethical principles, and privacy compliance (Madiega & Van de Pol, 2022). Notably, the Spanish sandbox is the first to be designed within the framework of the EU's AI Act, which includes provisions for pan-European coordinated AI sandboxes aimed at harmonizing regulatory efforts across member states (ANPD, 2023b; OECD, 2023).

In Latin America, Colombia, through its Colombian Superintendence of Industry and Commerce (Superintendencia de Industria y Comercio, SIC), has been operating an AI regulatory sandbox since 2021, with a particular focus on "privacy by design and by default" in AI projects. Launched with a public consultation in November 2020, this sandbox invited input from private companies, public authorities, and civil society (Muñoz et al., 2021) and was an inspiration for the ANPD's public consultation for the Brazilian AI sandbox (ANPD, 2023a). In 2021, Chile also presented its proposal for an AI regulatory sandbox, developed by the Chilean Ministry of Science, Technology, Knowledge, and Innovation (Ministerio de Ciencia, Tecnología, Conocimiento e Innovación, MSTC). This initiative is one of the provisions outlined in the new draft law on AI presented by the Chilean Government to Congress in 2024, with the stated goal of allowing startups to experiment with new technologies under regulatory supervision.

These international examples led the OECD (2023) to recommend that all countries adopt regulatory sandboxes to promote the development of responsible AI systems, underscoring their strategic importance in governing emerging technologies.

Forging a National Model: The Brazilian AI Regulatory Sandbox

In 2023, the ANPD initiated the development of a pilot regulatory sandbox for artificial intelligence (AI), pursuant to the provisions of Bill No. 2338/2023. Although this bill remains pending, the ANPD based its actions on existing legal frameworks, including the Brazilian Startups Legal Framework, which permits the creation of regulatory sandboxes, and the LGPD, which obliges the ANPD to establish guidelines for technologies involving personal data processing and to uphold the rights to revision and explainability of automated decision making.

To further this initiative, the ANPD (2023a) published a technical report comparing international AI regulatory sandboxes based on the experiences of Norway, the UK, Singapore, and Colombia. Drawing from these experiences, the ANPD defined the Brazilian sandbox as "a collaborative experiment between the regulator, the regulated entity, and other stakeholders, with the aim of testing innovations against the regulatory framework, using a structured methodology" (ANPD, 2023b, p. 11). The goals of this sandbox were defined as promoting algorithmic transparency, supporting responsible AI innovation, fostering multi-stakeholder collaboration on ethical principles, and ensuring human oversight of high-risk systems, with a focus on machine learning and generative AI (ANPD, 2023b).

In October 2023, ANPD (2023b) published a public call for input from society on the AI sandbox proposal through the Brazilian government's participatory platform, "Plataforma Brasil Participativo," which was open for one month. As part of the consultation, ANPD also hosted a webinar with representatives from Brazilian business, academia and civil society.

The consultation, which included open-ended questions, focused on five core dimensions of the sandbox: (1) its technological and (2) legal scope, (3) the transparency of the sandbox, (4) the structure of the sandbox and (5) stakeholder engagement (ANPD, 2023b). Participants were asked to discuss the selection criteria for participants, the role of key legal principles such as algorithmic transparency, and proposed methods for communicating the results of the sandbox to the public. Input was also sought on the duration of the sandbox, regulatory flexibility, and stakeholder participation mechanisms.

Mapping Stakeholders in Brazil's AI Sandbox Consultation

This study examined all positions available online regarding the Brazilian Sandbox Consultation on AI regulation, collected through internet search engines in October 2024 and through the complete transcription of the webinar organized by the ANPD. The searches were conducted both in the general web and in the database of the Brazil Participativo platform. Unfortunately, it was not possible to obtain, from any official channel of the Brazilian state, information on the total number of participants in the consultation, nor confirmation that the proposals available online reflected all contributions submitted.

As a result, 55 pages of written submissions and one hour and 34 minutes of oral contributions made during the consultation webinar were analyzed, encompassing the positions of seven entities in Brazilian civil society: the ABES, the ITS, the Economic Regulation Laboratory at the Rio de Janeiro State University (Laboratório de Regulação Econômica da Universidade do Estado do Rio de Janeiro, UERJ Reg), the Homeless Workers Movement (Movimento dos Trabalhadores Sem Teto, MTST), the Mozilla Foundation, the Laboratory of Public Policy and the Internet

(Laboratório de Políticas Públicas e Internet, LAPIN), and the Rights in Network Coalition (Coalizão Direitos na Rede, 2023). It is noteworthy that ITS and UERJ Reg submitted a joint contribution, while the Rights in Network Coalition and LAPIN were also represented jointly during the webinar, although LAPIN additionally provided its own autonomous submission.

ABES represents companies in Brazil's software and IT services sector and coordinates the Coalition for Innovation and Responsibility in Artificial Intelligence (CIRA), which includes associations tied to platforms like Uber, 99, and iFood. ITS, a nonprofit research institute founded by Ronaldo Lemos, a key figure in drafting Brazil's Internet Civil Rights Framework, focuses on ensuring equitable technological benefits for Brazil and the Global South and collaborated with UERJ Reg, a research group at Rio de Janeiro State University, in the sandbox consultation.

LAPIN operates as an independent center addressing social, ethical, and legal challenges of digital technologies, conducting research and advocating public policies. It is part of the Rights in Network Coalition, a group of over 50 organizations promoting digital rights, actively contributing to debates on Brazilian policymaking. The Mozilla Foundation, a non-profit organization known for free software projects such as Firefox, contributed to the consultation with insights on privacy and algorithmic racism through its Brazilian fellow Tarcízio Silva. The MTST, a Brazilian social movement advocating for housing and urban reform, has extended its efforts to include technology. Its tech collective, comprising tech workers, advocates for popular digital sovereignty and gained recognition with a member's appointment to the CNPD in 2024.

Society Speaks: Brazilian Perspectives on the AI Regulatory Sandbox

The positions expressed by civil society during the consultation on Brazil's regulatory sandbox for Artificial Intelligence (AI) converge on recognizing the importance of this methodology for governing technological development in the country. Nevertheless, a tension emerges between those actors who place greater emphasis on safeguarding rights and subordinating the sandbox to broader social policy goals, and those primarily concerned with competitiveness, legal certainty, and avoiding what they regard as excessive regulatory barriers for private developers.

What Should Be the Scope of the Brazilian Regulatory Sandbox(es) for AI?

Civil society proposals reveal two key points of contention related to the sandbox's scope and objectives, mirroring broader controversies over AI regulation in Brazil. The first relates to the current legal competencies of the ANPD in AI governance; the second questions whether regulation in general—and the sandbox in particular—should prioritize socially relevant criteria, such as impact on vulnerable populations and support for national companies, or whether it ought to remain as inclusive as possible without imposing significant barriers on private actors.

According to the ABES (2023), the LAPIN (2023), and the ITS in partnership with the UERJ Reg (2023), the existing legal framework grants ANPD supervisory powers over the use of personal data, thus limiting the sandbox to AI systems that explicitly involve this type of data processing. Conversely, the Mozilla Foundation (2023) and the MTST (2023) advocate for a broader scope, allowing any AI system that might—even indirectly—affect fundamental rights to enter the sandbox.

Within this rights-focused perspective, MTST (2023), LAPIN (2023), and Mozilla (2023) stress the importance of acknowledging Brazil's status as a developing country with significant socio-economic disparities. Drawing on the Colombian sandbox experience, they suggest that the sandbox should concentrate on specific national priorities and highlight a system's social impact. This may entail evaluating AI applications that disproportionately affect vulnerable populations—such as banking score models or technologies frequently used by marginalized groups (MTST, 2023)—or systems designed to mitigate disparities between urban and rural regions (LAPIN, 2023).

Mozilla (2023) also proposes prioritizing Brazilian companies to promote domestic innovation, while LAPIN (2023) argues that only companies with a local office in the country should be eligible. Both LAPIN (2023) and the Rights in Network Coalition (2023) emphasize the importance of accommodating independent entrepreneurs and autonomous developers. Meanwhile, ITS and UERJ Reg (2023) point out that the legislation does not impose formal restrictions on different types of companies, and ABES (2023) warns that overly detailed participation requirements may create prohibitive hurdles and discourage voluntary enrollment by companies. Additionally, MTST (2023) expands eligibility to include social movements, government agencies, and academic institutions interested in testing AI solutions.

Other proposed selection criteria vary in specificity. LAPIN (2023) requires evidence of prior experience, sufficient financial resources, and a minimally viable product. MTST (2023) calls for the inclusion of business models to assess market share, while ITS and UERJ Reg. (2023) argue for consideration of the transformative potential of the models, the severity of potential negative impacts, and the ability to respond to review or transparency requests. LAPIN (2023) similarly emphasizes that sandbox participants must adhere to privacy-by-design principles, justify why sandbox-driven regulatory flexibility is necessary, and explain how the system will benefit users.

From a technical standpoint, some proposals favor focusing on machine learning and generative AI due to their heightened risk of infringing fundamental rights (LAPIN, 2023), whereas ABES (2023) advocates expanding the scope to cognitive computing and neural networks, thus spurring research in less-developed fields. While most organizations support including both emerging and established technologies, LAPIN (2023) favors prioritizing AI still in its design stage, which presumably stands to benefit more from supervision and support.

Several actors also recommend creating thematic or sector-specific sandboxes to reflect Brazil's diversity in AI applications (ABES, 2023; LAPIN, 2023; MTST, 2023; Mozilla, 2023; ITS and UERJ Reg., 2023). For instance, this might entail initiatives for healthcare or education (LAPIN, 2023; ABES, 2023), sandboxes oriented by the target audience (ITS and UERJ Reg., 2023), or limited to specific technologies (LAPIN, 2023). Geographic boundaries may be relevant as well, such as municipal-level sandboxes (ITS and UERJ Reg., 2023). Opinions diverge over ANPD's precise role in establishing sector-specific sandboxes: some suggest that dedicated authorities (e.g., in health or agriculture) could lead their own programs (ABES, 2023), whereas others argue that ANPD's data protection expertise should enable it to run several sandboxes in partnership with other agencies (LAPIN, 2023; ITS and UERJ Reg., 2023).

The Battle for Algorithmic Transparency

Algorithmic transparency is widely seen as a pivotal—and contested—theme. On one side are those favoring enhanced mechanisms for explicability and review of automated decisions (LAPIN, 2023; MTST, 2023;

Mozilla, 2023), while on the other are actors who prioritize competitiveness and trade secrecy (ABES, 2023; ITS and UERJ Reg., 2023). Some organizations underscore those rights violations—such as copyright, privacy, or discrimination—already occur (Mozilla, 2023) and demand stricter rules. Others allude to "potential risks" and to ethical principles, warning that "excessive transparency" might deter companies (ABES, 2023).

For ABES (2023), it is crucial to adopt a balanced interpretation of transparency that does not impede AI innovations. The association therefore opposes making dataset disclosure a binding requirement, arguing that it could jeopardize industrial secrecy. MTST (2023) counters that the main risks stem from how model features shape automated decisions, particularly when sensitive data (e.g., race, address, sexual orientation, or religion) are involved; such transparency would give individuals the ability to understand how these variables are weighed by the system. Mozilla (2023) similarly supports robust auditing procedures, whereas LAPIN (2023) and the Coalizão Direitos na Rede propose a risk matrix to ensure human oversight of automated decisions and appeals to a human authority, especially in high-risk or irreversible scenarios.

ITS and UERJ Reg (2023) posits that the sandbox's core mission is methodological transparency, not necessarily broad algorithmic transparency, suggesting that disclosure measures be adapted to their context. Nonetheless, they emphasize that this environment can also serve as a testing ground for legal provisions under the LGPD that afford citizens the right to request reviews of automated decisions, thus revealing how far such obligations can extend without compromising trade secrets.

Between Innovation and Compliance: Mechanisms of Regulatory Flexibility

There is consensus that the ANPD should be able to suspend or adapt certain rules on a temporary basis so that potentially innovative technologies can be trialed without automatically violating the LGPD or proposed AI legislation. Even so, LAPIN (2023), MTST (2023), and ITS and UERJ Reg (2023) caution against exposing vulnerable groups—such as children or adolescents—to unacceptable risks during this transient deregulation. ITS and UERJ Reg. (2023) also stresses that the ANPD does not have the power to suspend or override statutes outside its purview, including treaties on human rights. ABES (2023), meanwhile, contends that the experimental nature of the sandbox opens the door for adopting a regulatory "trial-and-error" mindset typical of software development, thereby granting AI projects greater regulatory leeway.

However, in order to avoid legal uncertainty, the ANPD needs to clarify exactly which regulatory instruments would be suspended during each sandbox cycle, based on the prior consultation with participants (ITS and UERJ Reg., 2023). If flexible rules prove beneficial, the ANPD could grant permanent exemptions, provided they maintain market fairness and avoid favoring sandbox participants (LAPIN, 2023). MTST (2023) also suggests isolating feature engineering from business decisions to maintain methodological integrity during the sandbox, while LAPIN (2023) advocates involving ethics specialists to assess the use and risks of AI systems.

With regard to timing, most entities suggest an 18–24-month timeline for sandbox cycles, with flexibility depending on the complexity of technologies. MTST (2023), however, advocates a shorter schedule, citing the rapid pace of innovation. Another issue concerns the conclusion of the sandbox: if the final evaluation reveals improper use or unintentional consequences, participants may be prohibited from deploying the tested models in production unless they address the identified issues. It nonetheless remains unclear which risks would lead to immediate disqualification (ITS and UERJ Reg., 2023; LAPIN, 2023; MTST, 2023).

Shedding Light on AI Governance and Multi-stakeholder Participation

All civil society proposals identify public dissemination of sandbox results as a central requirement. Agents are thus encouraged or even obligated to produce algorithmic impact reports (Mozilla, 2023; LAPIN, 2023; MTST, 2023), with clear explanations of key factors employed by the models, along with potential effects. LAPIN (2023) similarly advocates data protection impact reports, spelling out risks and suggesting mitigation strategies, while ITS and UERJ Reg. (2023) urges each participant to detail which organizations contributed to product development and implementation. Both LAPIN (2023) and Mozilla (2023) call for documentation of the entire process, ensuring sufficient traceability for data and algorithmic auditing.

Proposals also address how the ANPD should communicate information about the sandbox. Several recommend publishing detailed reports on each project cycle, including final regulatory impact analyses (LAPIN,

2023; ITS and UERJ Reg., 2023), using language understandable to non-specialists (MTST, 2023; LAPIN, 2023) and holding public and online events (ITS and UERJ Reg., 2023). ABES (2023) stresses the importance of fostering digital literacy, while MTST (2023) further advocates offering programming courses so that the broader population can grasp how algorithms function in everyday decisions.

Following Brazil's tradition of multi-stakeholder digital governance, there is broad agreement on forming a multi-stakeholder committee of experts to supervise the sandbox. Suggested criteria include open calls for membership, parity between academia, government, civil society, and the private sector (MTST, 2023; ITS and UERJ Reg., 2023), and adequate representation of diverse knowledge areas, as well as gender, race, and region (LAPIN, 2023). This diversity not only enables more robust risk assessments (ABES, 2023; ITS and UERJ Reg., 2023) but also mitigates the risk of regulatory capture, whereby a regulatory agency becomes beholden to a specific stakeholder group (LAPIN, 2023). Nonetheless, ITS and UERJ Reg. (2023) clarifies that the committee's role is to assist the ANPD at key phases, such as project selection—possibly in a "blind" format based on predefined criteria—and the monitoring of outcomes, rather than supplant the ANPD's authority altogether.

In summary, while civil society representatives largely endorse the concept of a regulatory sandbox to stimulate innovation, protect fundamental rights, and foster legal certainty around AI, significant disagreements persist over the breadth of participation, the extent of algorithmic transparency, the interplay between flexibility and risk mitigation, and the precise contours of multi-stakeholder oversight. The sandbox thus emerges as an opportunity to experiment with novel AI solutions under regulatory guidance, but its ultimate success will depend on carefully balancing these diverse perspectives within Brazil's evolving legal and technological landscape.

Conclusion

This chapter has highlighted the multifaceted debates surrounding the establishment of a regulatory sandbox for AI in Brazil. This initiative serves as a microcosm of broader tensions over the role of public regulation in technological development. Discussions about the sandbox's scope and objectives have revealed fundamental disagreements about whether regulatory frameworks should primarily uphold rights or foster innovation, and how these goals can best be reconciled in practice.

A central debate concerns the interplay between general frameworks and sector-specific regulations. While Brazil's proposed comprehensive AI law aims to establish overarching guidelines, various stakeholders emphasize the need for targeted measures that address the unique risks and opportunities of particular technologies or application areas. Proposals for sector-specific sandboxes demonstrate an acute awareness of the complexity and diversity of AI systems. International experience suggests that tailored sandboxes can provide valuable insights (ANPD, 2023a), yet implementing such an approach in Brazil raises questions about the role of the ANPD.

Brazil is currently opting not to create a new, centralized AI governance authority. Instead, it is proposing a system in which various entities, coordinated by the ANPD, collectively oversee AI governance - potentially serving as a novel model for international debate. Nonetheless, a persistent challenge lies in clarifying the interrelation between the ANPD's existing data-protection mandate and its expanded responsibilities should the proposed Bill No. 2338/2023 be approved. This challenge is reflected in controversies over whether the sandbox should be limited to systems that process personal data, as in the Norwegian and British examples, or whether it should already anticipate broader AI-focused tasks—like the Spanish case. The same lack of clarity surfaces in discussions on whether the ANPD should exclusively conduct regulatory sandboxes on AI (in dialogue with sectoral agencies for specialized sandboxes) or merely supervise programs run by those agencies or even municipal entities.

Another recurring theme is the tension between Brazil's national priorities and international standards. While alignment with global frameworks is often seen as necessary for competitiveness (ABES, 2023), some stakeholders warn against uncritically adopting regulatory models designed in advanced economies (Mozilla, 2023; LAPIN, 2023). Instead, they advocate locally rooted approaches that consider regional needs and contexts - a position that aligns with broader discussions of digital sovereignty in the Global South and critiques of digital colonialism (Seto, 2024), even if these theoretical concepts do not appear explicitly in the analyzed corpus. Finding the right balance between local originality and international standard is a delicate task: minimal regulation without provisions for local industry could lead to over-dependence on powerful global players, while over-regulation could stifle national initiatives.

Further exploration of how such localized approaches can be developed and systematically integrated into global AI governance structures remains a key avenue for future research. Similarly, the role of South-South cooperation remains largely unexplored beyond an isolated mention by Mozilla's representative, Mr. Silva. Indeed, Brazil's membership in the BRICS Digital Partnership could become a more significant factor if integrated into the sandbox dialogue (Belli, 2021).

Moreover, the controversies underlying the sandbox approach reflect broader historical and social disputes in Latin America. Debates over the extent to which a looser or absent guarantee of rights can be tolerated in the name of development underscore the region's enduring historical patterns of conservative modernization that priorities economic growth over social welfare (Seto, 2024). These questions reflect traditional Latin American struggles between different conceptions of the role of the state, ranging from a more liberal vision focused on facilitating private enterprise to a more interventionist viewpoint advocating greater public direction of AI governance aimed at reducing social inequalities. Nevertheless, most proposals assume that AI development will be driven primarily by the private sector, with the notable exception of MTST (2023), which highlights the potential of cooperative and public institutions.

In this context, questions arise as to whether non-commercial or smaller-scale initiatives can meet the demanding eligibility criteria proposed by some actors, such as LAPIN (2023), or whether specialized sandboxes for startups and non-corporate ventures might be warranted. The chapter's discussions also highlight the importance of building transparency and accountability into complex technical development processes, but also the need to translate technical expertise for broader audiences. In this regard, the epistemological diversity advocated by LAPIN (2023) for the multi-sectoral committee is crucial, calling for broader representation from other fields of knowledge beyond IT and law, which has so far almost exclusively dominated Brazil's regulatory discourse on technology (Seto, 2021). Equally important is the inclusion of an intersectional perspective, recognizing that discriminatory biases often combine to compound and reinforce oppressions of race, gender, and class (Seto, 2024).

Ultimately, the Brazilian regulatory sandbox points to the possibility of fostering innovation in a structured environment while addressing the particular realities of the Global South. Its success will depend on how effectively these varied interests—ranging from competitiveness and innovation to the protection of fundamental rights and public engagement—are brought into alignment. In tackling these complexities, Brazil may well craft a distinctive regulatory model that meets international benchmarks yet remains responsive to the specific circumstances of Latin America.

References

Associação Brasileira das Empresas de Software. (2023, November 17). [Webinar presentation]. *Webinário—Sandbox Regulatório de Inteligência Artificial e Proteção de Dados.* Retrieved January 16, 2025, from https://www.youtube.com/watch?v=j4eQJXXz5I4

Autoridade Nacional de Proteção de Dados (ANPD). (2023a). *Sandbox Regulatório—Estudo Técnico.* ANPD. Retrieved January 16, 2025, from https://www.gov.br/anpd/pt-br

Autoridade Nacional de Proteção de Dados (ANPD). (2023b). *Consulta à Sociedade: Sandbox Regulatório de Inteligência Artificial e Proteção de Dados no Brasil.* ANPD. Retrieved January 16, 2025, from https://www.gov.br/participamaisbrasil/regulatory-sandbox-on-artificial-intelligence-and-data-protection-in-brazil

Belli, L. (2021). BRICS Countries to Build Digital Sovereignty. In L. Belli (Ed.), *CyberBRICS: Cybersecurity Regulations in the BRICS Countries* (p. 271). Springer.

Bundesministerium für Wirtschaft und Energie (BWE). (2019). *Regulatory Sandboxes as a Driver of Innovation.* Energy Transition—The German Energiewende. Retrieved January 16, 2025, from https://www.bmwk-energiewende.de/EWD/Redaktion/EN/Newsletter/2019/07/Meldung/topthema.html

Chinmayi, A. (2020). AI and the Global South: Designing for other worlds. In M. Dubber, F. Pasquale, & S. Das (Eds.), *Oxford Handbook of Ethics of AI* (Draft Chapter).

Coalizão Direitos na Rede. (2023, November 17). [Webinar presentation]. *Webinário—Sandbox Regulatório de Inteligência Artificial e Proteção de Dados.* Retrieved January 16, 2025, from https://www.youtube.com/watch?v=j4eQJXXz5I4

Conceição, L. H. M., & Perrone, C. (2022). The Brazilian proposed regulation of AI: Contextualization and perspectives. *MediaLaws.* Retrieved January 16, 2025, from https://www.medialaws.eu/the-brazilian-proposed-regulation-of-ai-contextualization-and-perspectives/

Crum, C. R., & Coglianese, C. (2024). Taking training seriously: Human guidance and management-based regulation of artificial intelligence. In *Proceedings of the IEEE International Symposium on AI Ethics and Governance.* https://doi.org/10.48550/arXiv.2402.08466

Derechos Digitales, Coding Rights, Fundación Karisma, Hiperderecho, IDEC, & R3D. (2024). *As vias regulatórias para a IA na América Latina: Compilação de estudos de caso no Brasil, México, Peru e Colômbia*. Derechos Digitales. Retrieved October 11, 2024, from https://www.alsur.lat/pt-br/relatorio/vias-regulatorias-para-ia-na-america-latina-compilacao-estudos-caso-no-brasil-mexico-peru

Gstrein, O. J., Haleem, N., & Zwitter, A. (2024). General-Purpose AI Regulation and the European Union AI Act. *Internet Policy Review, 13*(3), 1–26. https://doi.org/10.14763/2024.3.1790

Guha, N., Lawrence, C. M., Gailmard, L. A., Rodolfa, K. T., Surani, F., Bommasani, R., et al. (2024). AI Regulation has its Own Alignment Problem: The Technical and Institutional Feasibility of Disclosure, Registration, Licensing, and Auditing. *The George Washington Law Review, 92*(6), 1473–1552.

Hartmann, D., Laranjeira de Pereira, J. R., Streitbörger, C., & Berendt, B. (2024). Addressing the Regulatory Gap: Moving Towards an EU AI Audit Ecosystem Beyond the AI Act by Including Civil Society. *AI and Ethics*. https://doi.org/10.1007/s43681-024-00595-3

Institute for Technology and Society (ITS Rio), & Laboratory for Economic Regulation (UERJ Reg.). (2023). *Consulta Pública ANPD—Sandbox Regulatório*. Retrieved October 11, 2024, from https://itsrio.org/pt/publicacoes/consulta-publica-anpd-incidentes-de-seguranca-com-dados-pessoais/

Laboratório de Políticas Públicas e Internet (LAPIN). (2023). *Contribution to the Public Consultation of the National Data Protection Authority (ANPD) on the Regulatory Sandbox for Artificial Intelligence and Data Protection in Brazil*. Brasília: LAPIN. Retrieved January 16, 2025, from https://lapin.org.br/

Latin American Artificial Intelligence Index (ILIA). (2024). *Latin American Artificial Intelligence Index (ILIA) 2024*. CENIA. Retrieved October 11, 2024, from https://www.cepal.org/en/pressreleases/latin-american-artificial-intelligence-index-ilia-reconfirms-chile-brazil-and-uruguay

Lira, C. (2024). *Senado aprova marco regulatório da inteligência artificial: Entenda*. Fast Company Brasil. Retrieved January 16, 2025, from https://fastcompanybrasil.com/ia/senado-aprova-marco-regulatorio-da-inteligencia-artificial-entenda/

Machado, M., & Winter, E. (2023). Artificial Intelligence and Patents in Brazil: Overview on Patentability and Comparative Study on Patent Filings. *World Patent Information, 72*, 102177.

Madiega, T., & Van de Pol, A. L. (2022). *Artificial Intelligence Act and Regulatory Sandboxes*. European Parliamentary Research Service (EPRS). Retrieved January 16, 2025, from https://www.europarl.europa.eu/thinktank/en/document/EPRS_BRI(2022)733544_EN

Maslej, N., Fattorini, L., Brynjolfsson, E., Etchemendy, J., Ligett, K., Lyons, T., Manyika, J., Ngo, H., Niebles, J. C., Parli, V., Shoham, Y., Wald, R., Clark, J.,

& Perrault, R. (2023). *The AI Index 2023 Annual Report.* AI Index Steering Committee, Institute for Human-Centered AI, Stanford University.

Mendes, L. S., & Kira, B. (2023). The Road to Regulation of Artificial Intelligence: The Brazilian Experience. *Internet Policy Review.* Retrieved January 16, 2025, from https://policyreview.info/articles/news/road-regulation-artificial-intelligence-brazilian-experience/1737

Movimento dos Trabalhadores Sem Teto (MTST). (2023). *Contribuição para a Consulta Pública ANPD—Sandbox Regulatório.* Retrieved October 11, 2024, from https://nucleodetecnologia.com.br/docs/doc194428

Mozilla Foundation. (2023, November 17). [Webinar presentation]. *Webinário—Sandbox Regulatório de Inteligência Artificial e Proteção de Dados.* Retrieved January 16, 2025, from https://www.youtube.com/watch?v=j4eQJXXz5I4

Muñoz, V., Tamayo, E., & Guio, A. (2021). *The Colombian Case: Adopting Collaborative Governance as a Path for Implementing Ethical Artificial Intelligence.* Center for Technology and Society Studies (CETyS), University of San Andrés. Retrieved January 16, 2025, from http://hdl.handle.net/10908/18743

Organisation for Economic Co-operation and Development (OECD). (2023). *Regulatory Sandboxes in Artificial Intelligence.* OECD Digital Economy Papers, No. 356. OECD Publishing. Retrieved January 16, 2025, from https://www.oecd.org/termsandconditions

Ovanessoff, A., & Plastino, E. (2017). *How Artificial Intelligence can Accelerate Growth in South America.* Report, Accenture, BR, June.

Rights in Network Coalition. (2023, November 17). [Webinar presentation]. *Webinar—Regulatory Sandbox for Artificial Intelligence and Data Protection.* Retrieved January 16, 2025, from https://www.youtube.com/watch?v=j4eQJXXz5I4

Seto, K. S. (2021). Regulação de plataformas digitais: Uma revisão sistemática de literatura. *Revista Eletrônica Internacional de Economia Política da Informação, Comunicação e Cultura, 23*(3), 223–250. https://doi.org/10.54786/revista-eptic.v23i3.16341

Seto, K. S. (2024). Platform Sub-imperialism. *Big Data & Society, 11*(2), 1–12. https://doi.org/10.1177/20539517241249410

Wall, P. J., Saxena, D., & Brown, S. (2021). Artificial intelligence in the global south (AI4D): Potential and risks. In *Proceedings of the 1st Virtual Conference on Implications of Information and Digital Technologies for Development.*

Zanatta, R. A. F., & Rielli, M. (2024). *A construção da legislação de inteligência artificial no Brasil: Análise técnica do texto que será votado no Plenário do Senado Federal.* Retrieved January 16, 2025, from https://www.dataprivacybr.org/a-construcao-da-legislacao-de-inteligencia-artificial-no-brasil-analise-tecnica-do-texto-que-sera-votado-no-plenario-do-senado-federal/

PART II

Challenges and Promises of AI in the Public Sector

Ethical and Legal Dilemmas of National AI Policies in Latin American Countries

Rosa María Alonzo González

Introduction

The developing countries, as those in Latin America, are trying not to be left behind in the race to develop and incorporate artificial intelligence to improve their internal functioning and public services, and to face public problems that have been traditionally affecting the region: corruption, air pollution, high levels of criminality and recently, water shortage. They are fostering software such as chatbots and AI systems to improve urban mobility, along with smart devices as in the case of AI surveillance cameras and AI drones to inhibit crime. Recently, most Latin American countries have been working on framing AI national plans to set general policies to encourage advances in artificial intelligence in the forthcoming years.

R. M. Alonzo González (✉)
Facultad de Ciencias Humanas, Universidad Autónoma de Baja California, Mexicali, Baja California, Mexico
e-mail: rosa.alonzo@uabc.edu.mx

D. Ramírez Plascencia, R. M. Alonzo González (eds.), *Ethical and Legal Dilemmas of Artificial Intelligence in Latin America*,
https://doi.org/10.1007/978-3-031-86540-4_6

This work studies the main regional and national AI policies undertaken in Latin America. It describes their principles, main actions, and objectives. This work also criticizes their weaknesses and the legal and social barriers that could limit the impact of this policy in the region. As it will be possible to observe in this work, these actions may be insufficient, and some national and international factors may hinder the efforts. At a local level, Latin American governments lack basic infrastructure to support, not just the establishment of AI, but also high-speed internet connectivity. There are even some communities that are still in need of crucial public services such as water and electricity. It is also important to promote synergies among national and regional stakeholders to encourage investments to set up AI research laboratories and AI academic programs to foster not just new AI software and devices but also the fundamental digital competencies that help people incorporate those disruptive advances into their daily lives. Other issues, such as regulations and the set of ethical norms that guide these efforts must be built under a local-regional perspective. Countries must review actual legal frameworks and bring up-to-date outdated regulations about civil and criminal liability, intellectual property rights, and labor normatives, as well as promote ethical guidelines that protect users' and communities' rights. These guidelines, while observing general human rights principles, should be collectively created, focusing not only on foreign powers and international corporations' needs, but also on regional demands and long-lasting social vindications.

AI in Latin America: A North-South Technological Dilemma

As it has happened with other technologies that have been developed in rich countries and expanded across the globe, such as computers, the Internet, or social media, they are designed to serve political and economic interests that do not necessarily empathize with developing countries' needs. Most of the key regulations of big technological corporations (ALPHABET, META, or APPLE), despite their global reach, are rooted in the United States, therefore key legal issues about intellectual property law, free speech, and economics under which Facebook or Google operate mostly comply first with the US government's interests. The same applies to AI applications such as ChatGPT or the social media TikTok, whose developers must follow US and Chinese regulations respectively, and in some cases attend to their respective governments' legal impositions.

These tacit and explicit foreign principles materialize and determine how the algorithms that support AI applications operate (Mohamed et al., 2020). Besides helping users to accomplish a task, those applications keep users' data: personal information, political and commercial preferences, and hobbies. This information is employed by technological corporations not only to improve the applications and deliver a better service, but these data is also handled for marketing purposes, shared with local and foreign governments for surveillance, and potentially, to censor some contents and individuals. On some occasions, users' data could be eventually used to commit unlawful actions such as online frauds, or even to promote social and political instability and disinformation.

The North-South technological dilemma does not only stand challenges derived from regulation and political control, but also in ideological dominance through which contents, editorial lines, and perspectives are dictated to comply with cultural interests and agendas that do not necessarily express social needs and interests of marginalized communities and peripheral countries. This ideological control is also expressed in algorithms and content. Online services such as Internet search engines or chatbots provide biased outcomes when dealing with racial and gender controversial issues (Gibney, 2024; Esat, 2024). As it has traditionally happened with the Internet, ideological dominance also permeates disruptive technologies that are nourished from data already present online. The underrepresentation of racial minorities and women, therefore translated into artificial intelligence, particularly regarding the case of Generative AI. The rising predominance of AI in the economy and social spheres could eventually lead to the next digital divide, where rich countries get the most revenues from this advance. Besides, it has happened traditionally with other technologies and devices such as smartphones that depend on strategic raw materials such as cobalt and Lithium, usually extracted at low costs from Africa and Latin America. While AI developments are designed and created in rich countries as in the case of the USA or China, poor and developing nations stand as mere consumers of "prefabricated" applications, through which big corporations will continue operating a digital extractivist logic (Arroyo et al., 2019), profiting from private data and copyrighted material, relegating low-paid employments to southern countries, whose people work as data labelers feeding AI systems (Mishra, 2021).

It is precisely regarding the popularization of and expansion of AI that governments in Latin America are pondering the ethical principles, legal frameworks, and the best strategies to incorporate this disruptive technology into the public administration to improve internal performance and services. But also, to encourage the development of systems and applications that correspond to the national demands, avoiding external dependency on international tech corporations (Png, 2024). It is imperative to set strategies that support regional interchange and cooperation, but also establishing effective legal frameworks that protect and encourage private investments, particularly regarding the copyright of future developments.

AI Regional Policies

Incipient AI Policies in Latin America

Years ago, before the launch of ChatGPT and the popularization of Generative AI applications, international organizations started considering the future impact of artificial intelligence in the social and economic spheres. They were pondering which ethical and legal principles should regulate present and future developments, particularly. In 2019, the OECD (2019) introduced a series of key fundamental value-based principles, such as (a) Inclusive growth, Sustainable development, and well-being, (b) Human rights and democratic values, including fairness and privacy, (c) Transparency and explainability, (d) Robustness, security and safety, and (e) Accountability. These principles were updated in 2024 with recommendations for policymakers and to guide AI actors. In 2021, UNESCO (2021) delivered a statement about AI ethics that complemented previous OECD's principles with four core values and ten core principles laid out as a human-rights center. The international organization highlights the following issues: approach to (a) Safety and Security, (b) Responsibility and Accountability, (c) Right to Privacy and Data Protection, (d) Sustainability, (e) Fairness and Non-Discrimination. In 2024, these core values and principles were complemented with eleven key areas for policy actions and two practical methodologies to help their implementation.

It is possible to observe under both statements (OECD and UNESCO), that although the importance of AI, its development and expansion should be undertaken under a responsible, equitable, and secure form. Since public interests are confronting private corporations' goals, those ethical principles are the source of discussion and debate. Considering AI from the public policy spectrum, governments are not only focused on promoting technological change and economic development, but also are interested in surveillance, control, and regulation. However, these public considerations do not necessarily agree with big technological corporations' objetives, which are centered on obtaining revenues that compensate their investments. The actual AI political tug-of-war passed among those encountered visions: from one side, one that focused on tightening regulations and, on the other, those who praised more flexibility and self-regulation (The Economist, 2024).

In 2020, Fatima et al. (2020, p. 181) provided a general view of the most recurrent topics in the AI national strategic plans: (1) public sector functions (from healthcare, transportation, education, public safety, national security, among others), (2) industries (healthcare, financial, defense, and manufacture, etc.), (3) Data (Data exchange, data regulation, privacy, security), (4) Algorithms (Explainability, ethics, bias, trust), (5) Capacity Development (education, research and development, public policy-driven Innovations and financing) and (6) governance (regulation, risks, security, inequality and intellectual property rights protections, among others). As it is possible to observe, the government was not only pondering practical applications and key areas to implement AI but, there are also considerations about potential risks and the necessity to analyze actual regulations. Golubchikov and Thornbush (2020), for their part, summarized these issues at a local level. Those policies focused on AI and robotics employment (particularly drones) to improve the decision-making process in public administration, support education systems, and monitor water supply, urban mobility, and surveillance. However, as stated by the authors, while these policies provide technical characteristics to operate, there is little discussion about ethical and normative frameworks that will guide the functioning of these devices and developments, "Problems that are associated with cybersecurity and big data, including personal privacy—and ultimately democracy—need to be addressed" (Golubchikov & Thornbush, 2020, p. 6).

In the last four years, the government's interest in artificial intelligence in Latin America has risen significantly. Dozens of official reports and independent studies have been written around the potentialities, barriers, and risks of implementing this disruptive technology into the public sector. However, private corporations have been adopting AI into their production processes as well. By 2020, Latin American industries such as banks, airlines, and e-commerce companies such "Mercado Libre," a reference in the region, and Rappi, a delivery mobile application, were already using AI systems to improve their customer service and supply chain management (MIT Technology Review Insights, 2020). Brazil was among the first countries to establish a national plan to set up laboratories across the country to develop AI systems in cybersecurity and defense areas. Chile, for its part, established a coordinated action with civil organizations, to visualize how to set the best AI strategies in productive areas such as agriculture (MIT Technology Review Insights, 2020). In Argentina, the first efforts concentrated on establishing a National Artificial Intelligence Plan (NAIP). This action aimed to provide a legal framework that brings control to the national authorities about diverse AI issues such as intellectual property, liability, certifications, standards, etc. (TMG, 2020). Other national strategies have focused more on incorporation than regulation, as in the case of the Colombian government, whose National Development Plan (NDP) for 2018–2022 was praised for adopting disruptive technologies such as the case of AI to improve public services (TMG, 2020).

The outbreak of the COVID-19 pandemic in 2020 caused adverse health and economic impacts across the globe, but it also accelerated digital media adoption. Latin America was not the exception. Suddenly, governments, corporations, and the general people started virtualizing their daily activities. This promoted the development of digital competencies and the acquisition of smart devices, even amid an adverse economic context and in a region with huge social and economic disparities. People started using food delivery services and studied and worked remotely. Governments developed more online services, accessible from smartphones and tablets, and SMEs made efforts to put their services and products online (OECD, 2020). In 2021, more countries in the region counted or were working on an AI national strategy. However, the level of implementation differs among them. Some countries, such as Colombia and Uruguay, were already operative, others such as Brazil were being implemented, as well as Argentina and Chile, even though they already had an operative national plan. In the last cases, Mexico, Bolivia, and Ecuador have an AI national plan (Prudencio Ruiz, 2021).

An Actual Overview of AI Latin American Public Policies

The following passages provide a summary of the actual AI policies in Latin America. However, making an accurate analysis is complex since some national strategies are currently in development and other actions perhaps are no longer in operation or would have changed because some countries have had electoral changes in recent years. This was the case of an AI legal framework accepted in 2021 by the Brazilian House of Representatives, but that was not approved by the Senate. The document provided some principles and actions, but it did not set clear enforcement procedures (Baptista, 2023). This section focuses on analyzing Latin American AI policies undertaken in 2023 and 2024. The aim is not only to describe the main strategies but also to compare AI national plans and find common themes such as ethical principles and policies.

Argentina was one of the first Latin American countries to establish a strategic plan to stimulate AI adoption and research at a national level. In 2019, the president Mauricio Macri delivered The National Artificial Intelligence Plan to articulate the actions to employ the AI to promote economic growth, sustainability, and social inclusion (Presidencia de la Nación, 2019, pp. 14). The policy was also oriented to encourage the adoption of AI in the private and public sector, but this process must follow ethical criteria such as maximizing the people's life conditions and diminish potential risks. It also encouraged the development of technological competencies and talents at all educational levels and the establishment of an Ethical Committee that will supervise potential risks derived from the implementation of AI systems (pp. 96).

Recently, in 2023 the Argentinian government delivered the document "Recomendaciones para una Inteligencia artificial fiable" (Recommendations for reliable artificial intelligence), a brief guide that servers to highlights the ethical principles that will support the incorporation and development of artificial intelligence systems in the country: Human oversight and decision, transparency and explainability, responsibility and accountability, awareness and education and governance and adaptive multi-stakeholder collaboration (Secretaría de innovación pública, 2023, pp. 5). Additionally, there are other bills that have been considered right now in the congress such as the draft Law 2505-D-2023 titled "Marco legal para la regulación del desarrollo y uso de la Inteligencia Artificial" (Legal framework for the regulation of the development and use of artificial intelligence), this proposal focuses on instituting a legal framework to norm the research, use, and regulation of AI in Argentina. It likewise aims to promote innovation and progress in the AI field to the human benefit, avoiding risks and negative consequences (Art. 1). Article 4 gathered ethical principles already established in other countries' proposals and strategic plans such as "human dignity, privacy, transparency, responsibility and equity" (Congreso de la Nación Argentina, 2023). Article 9 also introduced the scheme of risks and proposed the creation of committees that will evaluate the potential perils of developing or using AI systems.

In Brazil, one of the first steps toward the implementation of AI in the public administrations could be routed to the Brazilian Digital Transformation Strategy in 2018 called (E-Digital). This plan encouraged the implementation of digital technologies to improve government performance, promote effectiveness in the private sector and digital inclusion in society (E-DIGITAL WORKGROUP, 2018). The E-Digital initiative also highlights the importance of developing robotics and artificial intelligence, but these actions should be followed along with an open debate about the protection of rights and the legal and ethical implications of the application of disruptive technologies such as AI and Internet of Things and robotics.

In 2021, the Brazilian Government erected a more specific strategy related with the development of AI. It was called EBIA. EBIA established three transverse axes—(1) Legislation, regulation and ethical use, (2) AI governance, and (3) International aspects—and six vertical axes—(1) Education, (2) Workforce and training, (3) RD&I and entrepreneurship, (4) Application in the productive sectors, (5) Application in the public power, and (6) Public security. Considering the first transverse axe, a subject of key importance in terms of this chapter, (1) Legislation, regulation and ethical: the document established the necessity of balancing the protection of basic rights such as personal data integrity and setting adequate legal frameworks to encourage technological development and legal certainty (Governo do Brasil, 2021, pp. 5).

Finally, the most recent action in the country has materialized in the bill 2338/23 (Senado Federal, 2023) that is now under consideration in the Brazilian Senate. In this case, it is not about a series of policies promoted by the government but concrete legal reforms to regulate the application and development of AI in Brazil. Art 2. Established ethical principles that will guide the expansion and operation of AI, here it is possible to highlight (a) human centrality, (b) respect to human rights and democratic values, (c) equality, and (d) privacy. Principles that concur with ethical frameworks supported by international organizations such as OECD and UNESCO.

Art. 4 provides key definitions including artificial intelligence "sistema computacional, com graus diferentes de autonomia, desenhado para inferir como atingir um dado conjunto de objetivos, utilizando abordagens baseadas em aprendizagem de máquina e/ou lógica e representação do conhecimento" (computational system, with different degrees of autonomy, designed to infer how to achieve a given set of objectives, using approaches based on machine learning and/or logic and knowledge representation). Chapter II, art. 5, 6, 7, 8, 9 and ten focus on providing a series of users' rights and securities against a potential abuse of AI systems that affects private individuals. This sphere of rights considers also the guarantee to be informed about the AI software operation (Art 5. Fracc. I), the right of privacy (Art 5. Fracc. VI), and no discrimination (Art 5. Fracc. IV). The bill pays attention moreover to the potential risks of using AI systems, it introduces in chapter III a scheme that categorizes risks according to their severity and establishes a series of steps to guarantee the public endorsement regarding the potential damages before the introduction of AI systems into the market.

Chile is another country that has been very active about AI and robotic regulation. Some of the first attempts to approach AI's normativity was in October 2021 (updated in 2024). The Chilean Ministry of Science, Technology, Knowledge and Innovation made public "The National Artificial Intelligence Policy." The document established three transversal principles: (1) an ethical and responsible use of AI, centered on the people, (2) AI at the service of sustainable development and (3) AI in international and multi-actor articulation. While the first two principles mainly focused on supporting an AI policy empathic with human rights and social and environmental sustainability. The last transversal principle promotes international cooperation and the synergies among multiple stakeholders. The National Artificial Intelligence Policy also considers three policy axes: (a) Enabling factors: such as talent and infrastructure development, (b) Adoption and development: actions to research, development, innovate and promote IA adoption, and finally (c) Ethic and Governance, an axe that embraces diverse activities to endorse regulation about intellectual property, the respect of human rights as the case of genre equality and non-discrimination and the respect of the environment. Among all these rights, it is possible to highlight the inclusion of a section dedicated to children and the young, this part endorses the promotion of AI use among minors, but from a responsible perspective that protects their privacy and keeps them safe from potential risks (Gobierno de Chile, 2024, pp. 45).

Along with the AI national plan, the legislative branch is working on a bill (15869-19) that regulates AI systems. It established a series of definitions such as AI system, user, provider, and developer. Then, it sets a series of guidelines regarding the operation of AI systems. It was established as the case of Brazilian regulation, a kind of scale that ponders the potential risk of employing AI software in determinate tasks. For example, art. 4 establishes some actions that abuse the user's vulnerability as "Unacceptable risk AI systems," this could be the case of people with mental disability. The bill also introduces the creation of the Comisión Nacional de Inteligencia Artificial (National Artificial Intelligence Commission) an agency that will validate and supervise the introduction of new AI systems into the market (Art. 06) (Cámara de diputadas y diputados de Chile, 2023).

Colombia has been very active formulating diverse documents to introduce AI into the private and public sphere in the country. In 2019, the government delivered the Política Nacional para la Transformación Digital y la Inteligencia Artificial (National Policy for Digital Transformation and Artificial Intelligence). This policy aims to generate social and economic value throughout the impulse of AI in the private and public sector. The document also established, following the OECD framework, 14 principles to guidelines to support the development of AI in Colombia. The seventh principle focuses on the ethical values, such as justice, freedom, non-discrimination, transparency, security, privacy, etc. (Gobierno de Colombia, 2019). A later document from 2021, "Ethical Framework for Artificial Intelligence in Colombia" (Gobierno de Colombia, 2021) would bring more details about these ethical principles. The objective of this ethical framework is to "provide recommendations and suggestions to Public Entities, to address the formulation and management of projects that include the use of artificial intelligence (AI)" (pp. 23). Supplementary to the ethical principles mentioned in 2019 policy (National Policy). The document served to provide details about these ethical codes. But it also recognized the "Human-in-the-loop" principle that establishes human supervision and control in AI processes (pp. 31). In addition, it mentions that all AI developments operated in Colombia should generate a clear social benefit (pp. 38). Finally, the document, following other countries' policies, introduces a system of risks associated with the implementation of AI, this according to the probability that the event occurs and its potential damages to the user.

In addition to these policies, the Colombian Senate is discussing right now the bill 059 (submitted in 2023) which "establishes the public policy guidelines for development, use and implementation of Artificial

Intelligence and other provisions are dictated" (Congreso de Colombia, 2023). This bill aims to set operational guidelines to develop AI systems following ethical principles already mentioned before. But it also set, as other cases, the creation of an office "La Comisión de Tratamiento de Datos y Desarrollos con Inteligencia Artificial" (The Commission for Data Processing and Developments with Artificial Intelligence) that will validate and supervise AI developments in the country (Art. 5).

Costa Rica's policies have centered on two main efforts. For one side, El Plan Nacional de Ciencia, Technología e Innovación 2022–2027 (Gobierno de Costa Rica, 2021). Published in 2021, praises for a responsible incorporation of AI in the Central American country through four main actions (Pag. 29): (a) the development of basic digital competences at school to support the use and research of AI, (b) the elaboration of a certification plan to capacitate professionals in the use of AI, (c) create public accessible AI systems, and (d) consolidated Costa Rica as the neuralgic center of AI development in Central America. The document also mentions the construction of a roadmap to establish an ethical framework to make responsible use of AI in the country. In recent years, there have been other actions (Asamblea Legislativa, 2024) focused on regulating AI in Costa Rica, such is the case of the Bill 23771. This legislative project aims to support the incorporation and expansion of AI in the nation following two main precepts: (a) ethical and secure use and (b) sustainable development. The proposal also established that its objective is "the protection and promotion of the dignity, human rights and well-being of the human person" (Art. 1). The document manifests a patent disposition to employ AI to improve the quality and accessibility of public services (Art. 2). The Costa Rican bill also mentions people's basic rights that should be respected by developments (Art. 4) such as human dignity, privacy, the right to free expression, and the principle of no-discrimination, among others. But it also has basic ethical principles (Art. 5): equity, responsibility, transparency, data protection, and AI developments should be focused on promoting social benefit, etc. Similar to the case of Chile, Costa Rica's bill also proposes the existence of an authority, in this case The Ministry of Science, Innovation, Technology and Telecommunications (MICITT), that will operate the registration, supervision and audit of artificial intelligence systems in the country.

The actions to establish ethical or legal frameworks to support the implementation of AI in Mexico have suffered a series of interruptions across the years. While Mexico was the first country to address the implementation of AI in March 2018, throughout the report "Towards an AI

Strategy in Mexico: Harnessing the AI Revolution" (Oxford Insights, 2018). This document launched among a political transition where Andres Manuel Lopez Obrador took possession of the Mexican presidency, proposed five main areas of interests where it is possible to highlight the following topic: (a) governance, government and public services; It recommended, among other strategic lines, to set a clear strategic direction and create a working group in the congress, (b) research and development; to create a national research center, (c) capacity, skills and education, develop tools for continued education in AI, and promote the development of AI competencies at school, (d) data and infrastructure: development training data to development AI applications, and protect personal privacy and (e) create a Mexican AI Ethics Council. However, this document did not have the impact during Lopez Obrador's presidency. It was until 2024 that some public and private efforts, throughout the transdisciplinary group ANIA -La Alianza Nacional de Inteligencia Artificial- The National Artificial Intelligence Alliance, delivered some studies that discuss AI policies and barriers to encourage the adoption of and development of artificial intelligence in Mexico. Considering the ethical aspects, ANIA, in their report "Proposal for the National Agenda for Artificial Intelligence for Mexico 2024–2030" (ANIA, 2024), suggests the set of ethical procedures that will guide the development of AI systems in the country, "Promote ethical and responsible management in the development, use, application and implementation of AI-based technologies" (pp. 24).

There have been other efforts in the legislative branch to set clear regulations of AI. In April 2024, Senator Ricardo Monreal (2024) presented the bill "Federal Law Regulating Artificial Intelligence." Among its main topics, Art. 2 established that AI systems must guarantee the protection of human rights. It also mentions in Art. 6 the confirmation of a national commission that will provide advice regarding further AI policies in the country. It also introduces, as in other countries' bills in the region, a scheme of potential risks of using AI systems (from low to unacceptable).

The first actions to encourage AI adoption in Peru could be rooted to the elaboration of the National Artificial Intelligence Strategy in 2021 (Gobierno del Perú, 2021). The main objective of this policy was to provide "axes, objectives, actions that promote the research, development and adoption of artificial intelligence, which helps create solutions to national problems based on AI" (pp. 51). Among those national problems, the information focuses on alleviating poverty, improving education quality, public health, providing clean water, and supporting a sustained urban development (pp. 87). The document also proposed six basic axes, among them was to promote AI competencies development, increase the economy using AI systems, create data infrastructure, adopt ethical guidelines, and promote international cooperation. (pp. 56). In 2023, the Peruvian government approved Law 31814 (Gobierno del Perú, 2023), a regulation that promotes the use of artificial intelligence in favor of the economic and social development of the country which established interest in endorsing AI to favor social and economic development, along with the improvement of public services, education, health, and security (Art. 2).

In Uruguay, the efforts to conform a national IA strategy could be dated to the document "Artificial Intelligence Strategy for Digital Government" whose objective was to boost the adoption of AI in the public administration. The document stated that AI should be used to empower people's capacities and promote the improvement of their life quality (AGESIC, 2020, pp. 10). In addition, it established a series of principles that should guide the implementation of AI in the government, such as the respect of human rights, transparency, privacy, security and responsibility, etc. (pp. 10). In 2024, the Uruguayan government delivered a new version of the AI national strategy, "National Artificial Intelligence Strategy Uruguay 2024–2030." This new document is more encompassing since it not only emphasizes on the public sector, but it also pays attention to the adoption of AI in the private sphere. The new strategy promotes the use of AI to increase economic growth, social inclusion, environmental sustainability, and strengthen the country's sovereignty (AGESIC, 2024, 06). Regarding the ethical aspect, the policy also establishes that "AI systems must be developed and used under ethical principles -privacy, freedom, non-discrimination, equality-, in a manner compatible with human dignity and respectful of human rights." (AGESIC, 2024, pp. 25). In August 2023, Senator Juan Sartori (2023) proposed a bill to regulate the artificial intelligence systems. This regulation provides operational guidelines to developers and users, particularly regarding the proper identification of a system that employs AI in its functioning (Art. 04) but also compelled users to publicly state that a multimedia element was made using AI (Art. 05), this to avoid deceiving content consumers.

Challenges, Risks, and Ethical Issues

The OECD (2019) recommendations for policymakers provide considerations to set legislative and regulatory frameworks, such as: (a) Investing in AI research and development, (b) Fostering an including AI-enabling ecosystem, (c) Shaping and enabling interoperability governance and policy environment for AI, (d) Building human capacity and preparing for labor market transformation, and (e) International co-operation for trustworthy AI. These recommendations frame the huge challenges that Latin America faces to ethically incorporate AI systems and its life cycle into its public policies.

Considering the ethical question, Brazil set in 2018 the E-Digital plan that encouraged the technological development of an efficient government and the promotion of digital inclusion, the respect of rights, and openned the debate about AI's legal and ethical implications. The EBIA strategy in 2021 established the 'protection of users' rights and a legal framework to encourage technological development. In addition, Brazil has under consideration the Bill (2338/23), an extensive legal frame that clarifies basic terms and provides users with protections from AI developers. It contains ethical principles that harmonize with international organizations (Human rights protections, privacy, human centrality, etc.). In Chile, the National Artificial Intelligence Policy (2021, updated 2024) endorses AI research, development, and adoption under specific ethical guidelines such as human centrality, sustainability, and the protection of individual rights. It also promotes regional integration and emphasizes securing the minor's experience when working with AI applications.

Since 2021, Costa Rica with The National Science, Technology, and Innovation Plan 2022–2027, has focused on AI competencies and systems development. On the other hand, the legislative project (Bill 23771) focuses on establishing ethical and legal principles to operate and develop AI systems. It also introduces the creation of a supervisor organism to validate the operation of AI software, as in the case of Chile and Brazil. Colombia and its National Policy for Digital Transformation and Artificial Intelligence (2019) mentions ethical principles that follow the OECD AI ethical framework. Then the Ethical Framework for Artificial Intelligence in Colombia (2021) enhances the details and description of these ethical principles. In addition, it mentions the human-in-the-loop code and introduces a system of AI operation risks. Nowadays in discussion, bill 059 (submitted in 2023) gives operational guidelines to develop AI software and establish an office to validate the AI systems that will operate in the country.

In Mexico, despite being the first country to address the implementation of AI in March 2018, throughout the report "Towards an AI Strategy in Mexico: Harnessing the AI Revolution" it was not until 2024 in the

"Proposal for the National Agenda for Artificial Intelligence for Mexico 2024–2030," that suggest the set of ethical procedures that could guide the development of AI systems in the country. Furthermore, the legislative branch presented the bill "Federal Law Regulating Artificial Intelligence" in April 2024. This regulation presents a scheme of AI risks, a consultative group creation, and guidelines to protect human rights. In 2021, Peru released The National Artificial Intelligence Strategy that provided guidelines to develop AI. In 2023, the country approved the Law N° 31814. It established main imperative areas where AI systems could improve the quality of public services or diminish social and economic problems.

Since 2019, Argentina with The National Artificial Intelligence Plan has encouraged the adoption of AI to diminish social inclusion and promote economic growth. This plan also recommends the establishment of guidelines to avoid potential risks. In 2023, Argentina introduced the ethical principles to adopt AI in the country, with recommendations to develop reliable artificial intelligence systems. Moreover, the draft of Law 2505-D-2023 introduces a legal framework to regulate the research, implementation, and use of AI systems. Lastly, in 2020, Uruguay with the "Artificial Intelligence Strategy for Digital Government," established ethical principles to incorporate AI in the public sector. In 2024, the Uruguayan government delivered a new version of the AI national strategy, "National Artificial Intelligence Strategy Uruguay 2024–2030" to enhance the AI policies in the public and private spheres of the country. It also provides general objectives and ethical principles regarding the employment of AI systems.

Conclusions

The analysis of national policies and legal bills developed in Latin America allows us to observe the difficulties of establishing a national framework that not only could guide the adoption and development of AI but also offers legal certainty to developers, investors, and users about the rights, liabilities, and potential risks derived from the operation of AI devices and systems. While most of the countries in Latin America have been working on general policies and regulations, the absence of secondary laws that provide technical rules and procedures to developers, public officers, and users is patent. Along with this lack, there are also imperatives to promote effective regional cooperative strategies that harmonize regulation and arrange collective investments.

Among the most challenging barriers to implementing AI in the region are the high costs of developing and implementing AI systems and to solve regulatory and security concerns. The first issue does not only imply the acquisition of infrastructure, but also human resources, such as AI developers, the acquisition of data, and the payment of software licensing, and in some public areas as in the case of security, the purchasing of sophisticated devices such as drones. Considering regulation and security, even though there have been some efforts to develop AI national strategies, it is important to expand those actions to renew outdated regulations, particularly regarding intellectual property law, criminal law, and other commerce regulations, etc. It is quite important to provide public officers with legal tools to protect AI developer's investments and user's rights from potential corporation and public abuses, as in the case of private data misuse that exposes users to the economic exploitation of their data and security vulnerabilities when using AI apps.

It is also important to unify and integrate these national policies from a regional and international perspective. A regional vision will provide national governments with legal and financial frameworks that allow them to cooperate and promote the development of AI laboratories, academic formation, private investments to build AI development centers and support regional working groups to raise their voice and impulse a regional agenda before international organizations such as OAS - Organization of American State and the United Nations.

References

Asamblea Legislativa. (2024). *Ley para regular la Inteligencia Artificial en Costa Rica.* Asamblea Legislativa República de Costa Rica.

AGESIC. (2020). *Estrategia de Inteligencia Artificial para el Gobierno Digital.* República Oriental del Uruguay.

AGESIC. (2024). *Estrategia Nacional de Inteligencia Artificial Uruguay 2024–2030.* República Oriental del Uruguay.

ANIA. (2024). *Propuesta de Agenda Nacional de la Inteligencia Artificial para México 2024–2030.* Alianza Nacional de Inteligencia Artificial.

Arroyo, L., Amjad, O., & Murillo, D. (2019). *My Data, My Rules. From Data Extractivism to Digital Empowerment.* La Caixa/ACUP.

Baptista, D. (2023, July 10). ChatGPT drafts law as Latin America seeks to regulate AI. *Context.* Retrieved July 22, 2023, from https://www.context.news/ai/chatgpt-drafts-law-as-latin-america-seeks-to-regulate-ai

Cámara de diputadas y diputados de Chile. (2023). Proyecto de ley que regula los sistemas de inteligencia artificial, la robótica y las tecnologías conexas, en sus distintos ámbitos de aplicación.

Congreso de Colombia. (2023). *PL 059-23 Inteligencia Artificial.* Pub. L. No. 059 (23).

Congreso de la Nación Argentina. (2023). Marco legal para la regulación del desarrollo y uso de la Inteligencia Artificial. Pub. L. No. 2505-D-2023.

E-DIGITAL WORKGROUP. (2018). *Digital Transformation Strategy (e-Digital).* Governo do Brasil. Retrieved December 5, 2024, from https://www.gov.br/mcti/pt-br/centrais-de-conteudo/comunicados-mcti/estrategia-digital-brasileira/digitalstrategy.pdf

Esat, S. (2024). From Search Engines to Debit Cards: How Technology Reinforces Social Inequalities. *Essex Student Journal, 15*(S1). https://doi.org/10.5526/esj.348

Fatima, S., Desouza, K. C., & Dawson, G. S. (2020). National Strategic Artificial Intelligence Plans: A Multi-dimensional Analysis. *Economic Analysis and Policy, 67*, 178–194. https://doi.org/10.1016/j.eap.2020.07.008

Gibney, E. (2024). Chatbot AI Makes Racist Judgements on the Basis of Dialect. *Nature, 627*(8004), 476–477. https://doi.org/10.1038/d41586-024-00779-1

Gobierno de Chile. (2024). *Política Nacional de IA.* Gobierno de Chile.

Gobierno de Colombia. (2019). *Política Nacional para la Transformación Digital y la Inteligencia Artificial.* Gobierno de Colombia.

Gobierno de Colombia. (2021). *Marco Ético para la Inteligencia Artificial en Colombia.* Departamento Administrativo de la Presidencia de la República.

Gobierno de Costa Rica. (2021). *Plan Nacional de Ciencia, Tecnología e Innovación 2022–2027.* Ministerio de Ciencia, Innovación, Tecnología y Telecomunicaciones.

Gobierno del Perú. (2021). *Estrategia Nacional de Inteligencia Artificial 2021–2026.* Secretaría de Gobierno y Transformación Digital.

Gobierno del Perú. (2023). Ley N.° 31814.

Golubchikov, O., & Thornbush, M. (2020). Artificial Intelligence and Robotics in Smart City Strategies and Planned Smart Development. *Smart Cities, 3*(4), 1133–1144. https://doi.org/10.3390/smartcities3040056

Governo do Brasil. (2021). *EBIA Strategy.* Governo do Brasil. Retrieved December 5, 2024, from https://www.gov.br/mcti/pt-br/acompanhe-o-mcti/transformacaodigital/arquivosinteligenciaartificial/ebia-summary_brazilian_4-979_2021.pdf

Juan Sartori. (2023). *Sistemas de Inteligencia Artificial Regulación.* Retrieved from https://parlamento.gub.uy/documentosyleyes/documentos/versiones-taquigraficas/senadores/49/1737/0/PDF

Mishra, S. (2021, March 9). Is AI Deepening the Divide Between the Global North and South? *Newsweek*. Retrieved April 19, 2024, from https://www.newsweek.com/ai-deepening-divide-between-global-north-south-opinion-1574141

MIT Technology Review Insights. (2020). *The Global AI Agenda: Latin America*. MIT. Retrieved August 2, 2023, from https://mittrinsights.s3.amazonaws.com/AIagenda2020/LatAmAIagenda.pdf

Mohamed, S., Png, M.-T., & Isaac, W. (2020). Decolonial AI: Decolonial Theory as Sociotechnical Foresight in Artificial Intelligence. *Philosophy & Technology, 33*(4), 659–684. https://doi.org/10.1007/s13347-020-00405-8

OECD. (2019). *AI Principles Overview*. OECD. Retrieved November 14, 2024, from https://oecd.ai/en/principles

OECD. (2020). *Latin American Economic Outlook 2020: Digital Transformation for Building Back Better*. OECD. https://doi.org/10.1787/e6e864fb-en

Oxford Insights. (2018). *Towards an AI Strategy in Mexico: Harnessing the AI Revolution*. British Embassy in Mexico.

Png, M.-T. (2024). At The Tensions of South and North: Critical Roles of Global South Stakeholders in AI Governance. In J. B. Bullock, Y.-C. Chen, J. Himmelreich, V. M. Hudson, A. Korinek, M. M. Young, & B. Zhang (Eds.), *The Oxford Handbook of AI Governance* (pp. 981–1014). Oxford University Press. https://doi.org/10.1093/oxfordhb/9780197579329.013.57

Presidencia de la Nación. (2019). *Plan Nacional de Inteligencia Artificial (ARGENIA)*. Gobierno de Argentina.

Prudencio Ruiz, K. (2021). *Estrategias de Inteligencia Artificial en América Latina*. ILDA.

Ricardo Monreal. (2024). Iniciativa con proyecto de decreto por el que se expide Ley Federal que Regula la Inteligencia Artificial.

Secretaría de innovación pública. (2023). *Recomendaciones para una Inteligencia artificial fiable*. Gobierno de Argentina.

Senado Federal. (2023). *Projeto de Lei N° 2338*. Retrieved from https://legis.senado.leg.br/sdleg-getter/documento?dm=9347622&ts=1733409396734&disposition=inline

The Economist. (2024, May). AI firms mustn't govern themselves, say ex-members of OpenAI's board. *The Economist*. Retrieved May 31, 2024, from https://www.economist.com/by-invitation/2024/05/26/ai-firms-mustnt-govern-themselves-say-ex-members-of-openais-board

TMG. (2020). *Overview of AI Policies and Developments in Latin America*. Telecommunications Management Group. Retrieved July 23, 2023, from https://www.tmgtelecom.com/wp-content/uploads/2020/03/TMG-Report-on-Overview-of-AI-Policies-and-Developments-in-Latin-America.pdf

UNESCO. (2021). *Ethics of Artificial Intelligence*. UNESCO. Retrieved November 14, 2024, from https://www.unesco.org/en/artificial-intelligence/recommendation-ethics

The Use of Biometrics and AI for Border Control and Its Impact on Human Rights

Jezabel Pérez Yáñez and Natalia Brzezinski Ramírez

Introduction

The robust capabilities of artificial intelligence to resemble human reasoning have allowed users to incorporate these systems beyond recreational issues. The presence of AI is growing in educational, economic, and public administration sectors. Governments are using it to improve public services and efficient communication channels with the citizens. By now, it is possible to observe how chatbots deliver information and manage procedures through smartphones. Governments are also using AI to combat criminality, as the case of COMPAS (Correctional Offender Management Profiling for Alternative Sanctions), a software employed in the United States to predict criminal recidivism or the employment of AI cameras that uses facial recognition to track suspects in real time. Other systems are

J. Pérez Yáñez (✉)
University of Guadalajara, Guadalajara, Jalisco, Mexico

N. Brzezinski Ramírez
University of Louisiana at Lafayette, Lafayette, LA, USA

D. Ramírez Plascencia, R. M. Alonzo González (eds.), *Ethical and Legal Dilemmas of Artificial Intelligence in Latin America*,
https://doi.org/10.1007/978-3-031-86540-4_7

already operating in national security, as novel defensive systems depend on AI capabilities to increase their efficacy against missile and drone swarms (Tucker, 2024). However, the growing presence of AI in local and homeland security is not immune to critics that question the risks of data breaches, the potential unauthorized access and control to the systems, and fatal errors when innocents are considered as targets under a determined algorithm preset (Dawes, 2021). Other questions have appeared in recent years because of the incorporation of AI systems along with biometric technology to manage migration flows in Europe and America.

Ideally, the use of technology can improve the speed and efficiency of bureaucratic paperwork by automating many of the procedures. In Roraima, at the Venezuelan Brazilian border, for example, IOM (International Organization for Migration) operates the DTM (Displacement Tracking Matrix) to collect data for managing aid, map migrant routes, and help support migrants during their journey. Notwithstanding, the methods employed to obtain the information are the source of suspicion because there are no guarantees about the integrity and preservation of migrants' data and the potential political uses of that information (Koch, 2023). Some testimonies from migrants who have passed through the Venezuela-Brazil border express concerns about the future use of their personal data. But, because of the urgency to leave their country and to enter Brazil legally, they are compelled to trust the organizations that were carrying out the process (Camargo & Alencar, 2024). In the United States, the destination of most of the migration flows in Latin America, the employment of technology in immigration procedures has generated complaints during the collection of biometric data. The capturing of personal information could be illegal since there are no guarantees about the protection of people's privacy or the potential misuse of this information to create biased negative profiles that affect the outcome of asylum solicitors (France 24, 2023). Faced with the growing problem of the transgression of the human rights of migrants, diverse organizations such as the Coalition Against Detention have been critical about the capture of migrant biometric data and the potential employment of AI to manage that information (Gottardo et al., 2024). This chapter focuses on the legal aspects of using biometric technology and AI to manage migration. This work focuses mainly on two cases: (a) the employment of biometric technology and AI by the DHS in the United States, and (b) the capturing of personal data of migrants in the Brazil-Venezuela border. It particularly analyzes how these technologies operate, the legal frameworks' violations and the actions undertaken by NGO organizations toward the technologization of border control.

Migration Crises, Mediated Refugees, and the Technologization of the Border Control

The actual migration flows in Latin America have been triggered by diverse factors such as economic depression, political repression, and high levels of insecurity, along with the negative effects caused by the 2020 pandemic. By now, there are thousands of people from Haiti, Venezuela, and Cuba, along with the traditional flows from Central America and Mexico, that are waiting in ill-equipped shelters at the US-Mexican border for the possibility to enter the United States. Nevertheless, the rising administrative barriers imposed by the US government and the imminent arrival of Donald Trump to the presidency for a second mandate in 2025 are limiting migrants' options. The presence of migrants in local communities has become the source of conflicts with locals. Disputes are exacerbated by biased media coverture and fake news. Sometimes the newcomers are considered a threat, as people who are "stealing jobs" or taking advantage of social welfare, but also as the source of criminality and a potential cause of contagion as seen during the 2020 pandemic (Ramírez Plascencia, 2024). This negative vision is also palpable on social media, since people have passed to associate migration with altruism to menace and burden (Avraamidou & Eftychiou, 2022). These negative portrayals have served as an excuse for federal governments to raise the migration controls, not only in the US but also across the continent. In 2022, for example, the US government ended the program "Stay in Mexico" that allowed migrants to apply at the US-Mexican border. By now, all applicants must ask for asylum status in their homelands. In 2024 the Biden administration allowed authorities to perform direct deportations to those people who were able to comply with the asylum requirements (France 24, 2024). Before the infeasibility of arriving in The United States, migrants are facing the dilemma of asking for asylum and establishing themselves in Mexico or returning to their countries. Other countries in Latin America, as in the case of Mexico, have toughened their immigration measures such as the imposition of visas for Cuban or Venezuelan citizens. The media has echoed the frequent complaints about the discriminatory treatment of

Mexican authorities toward Latin American travelers at airports who many times are denied entrance to the country.

Recent migration crises across the globe have also been shaped by the rising presence of information technology; migration is now mediated. Nedelcu and Soysüren (2022) mention two aspects derived from this process. On one hand, technology is an ally to refugees by helping them with plans of allocating. It provides them with valuable information related to the logistic and legal aspects, along with advice to diminish the potential perils. But on the other hand, the technological advances on digitalization and automatization have expanded government capacities to surveil and control people's activities online and in person. As digital media becomes essential for people, their interactions and digital recordings on the Internet have become key assets not only for corporations that commercialize this information, but also to governments. Regarding Latin American migrants, for example, the access to the Internet and the use of smart gadgets increase their opportunities to arrive into the US Mexican border, but the same technologies expose them to be surveilled and tracked by the migration authorities, and the digital private records in their smartphones are customarily used as proof to deny them entrance at airports and border controls.

Along with the rise of administrative controls, there is a growing employment of information technology at airports and borders. In fact, since the 9/11 terrorist attacks, there has been a continuous technologization of the border security in The United States (Felbab-Brown, 2021). One of the earliest actions in 2003 was the creation of the Customs and Border Protection (CBP), a federal agency combined with the previous U.S. immigration Custom Service. The agency has introduced diverse technologies to expand its surveillance capabilities, such as drones, biometrics, and AI to control the influx of merchandise and people coming into the US. The introduction of Artificial Intelligence (AI) offered the CBP new possibilities to carry out border control, such as video and image amendment, that implies building object detection models to automatically identify targets (U.S. Customs and Border Protection, 2023). The employment of disruptive technologies, as the case of AI and drones in national borders, not only in the US, but also in other regions such as Europe, has consolidated as what some authors call the rise of Intelligent Borders or smart borders (Korkmaz, 2021) where the intervention of humans in strategic security actions are less determinant, and the role of AI systems are becoming more imperative. However, for migrants that are trying to enter the US, the application of those technologies constitutes a threat to their private sphere, since there are no guarantees of the preservation of data or legal safeguards that protect them from filtrations and misuse of the information.

Smart borders are not exclusive to the United States or Europe. Its implementation could be observed in other Latin American countries (Alarcón et al., 2024), as in the Brazilian Venezuelan border. In 2021, the Brazilian Agency for Industrial Development (ABDI) initiated the modernization of border control with the project "Fronteira Tech," which includes the installation of intelligent lights, facial recognition systems, PTZ cameras with motion sensors, drones with thermal cameras, and a center for storage and processing of data and images (Lemos, 2021). Because of the migration crises, the specific situation of Brazil, which shares part of its northern border with Venezuela, impulses the foundation of the Acolhida Operation, which seeks the voluntary, safe, orderly and free relocation for Venezuelan migrants and refugees in vulnerable situations from the border of Roraima to other cities in Brazil (ACNUR, 2021). In this instance, migrants who wish to cross the border must supply biometric data, this activity is carried out by the Federal Police and the Federal Public Defender's Office Police and UN humanitarian organizations. These humanitarian organizations collect this data for the PRIMES system (Population Registration and Identity Management Ecosystem). This system is employed, among other purposes, to provide people with a digital identity to access different social services, while state agencies use the personal and socio-economic data for the relocation of migrants.

AI, Biometrics, and Migration: Legal and Ethical Concerns

The growing technologization of border control, including drones, biometrics and AI, have been the source of concerns among scholars and practitioners because of the potential violation toward migrant human rights. This situation is even worse for those people living in social vulnerability that were compelled to leave their homeland, escaping political repression, violence, or economic precarity, and are travelling across the continent trying to apply for asylum. It has been documented that Immigration and Customs Enforcement (ICE) routinely request personal information to online services and social media platforms without a court order (Bhuiyan, 2023). Other data corporations such as LexisNexis and Motorola Vigilant's Solution facilitate the tracking of people (Alarcón et al., 2024). The establishment of recent policies in the US regarding migration such as the imposition of the mobile application CBP One, that manages appointments to US migration authorities, implies other vulnerabilities such as to widen the digital divide among migrants, limiting the access to those people who do not have the digital competencies and infrastructure to use the application, or the lack of official language proficiency as the apps' interphase is only available in English, Spanish, Haitian Creole, Portuguese and Russian, and does not consider Indigenous languages. There do exist non-profit organizations that provide assistance to migrants in their application process, particularly due to the language barrier, but these organizations may be poorly staffed and are at times unable to legally interfere in the process. Without any alternative to arrange an appointment with migration authorities, some of the applicants are "left to their fate" exposed to digital scams by organized crime which exploits applicants (Dib & Deslandes, 2023).

The operation of CBP One implies the supply of personal and biometric data of users, including name, date of birth, facial images, and geolocation among other sensitive information. However, as some authors state, the US government shares this data with other federal agencies without the migrants' consent (Dávila et al., 2024). The Department of Homeland Security is also implementing Artificial Intelligence (AI) to make decisions about public safety menaces and the location and deportations of migrants. The use and storage of migrants' data is becoming the source of criticism, since the owners did not provide an authorization or were not aware of the implication of providing their data. There are also implications when the responsibility of assigning relief or guaranteeing an asylum request depends on algorithms already pre-designed, and sometimes biased (Saunders, 2023; Kinchin, 2024).

The establishment of AI for migration control, besides their critics and controversies, contrasts with the customary international regulation that protects migrants' human rights and forbids discrimination. Art. 17 of the International Covenant on Civil and Political Rights (UN, 1976) forbids "arbitrary or unlawful interference with his privacy." National authorities across the migratory routes in Latin America usually omit this right when forcing individuals to provide their biometric data and personal information. Other issues are related with how data is employed for facial recognition and the profile of potential suspects using AI which compromises the international regulation about discrimination as established in Article 02 of the International Convention on the Elimination of All Forms of Racial Discrimination (UN, 1965), which compelled all signed states not to "act or practice of racial discrimination against persons, groups of persons or institutions." These legal safeguards are also established in the US regulation. The 4th amendment of the US constitution sets "The right of the people to be secure in their persons, houses, papers, and effects, against unreasonable searches and seizures" (Congress.gov, 2025). Computer Matching and Privacy Protection Act (US Congress, 1988), besides its temporality, also ensures privacy, integrity, and verification of data in computer matching (which may be extensive to electronic devices). Recently, in 2024 US President Biden (Biden, 2024) signed a memorandum to allow national security agencies to employ advanced AI systems. While the official document specifies the protection of human rights such as privacy and safety, the patent lacks safeguards and accountability mechanisms that protect individuals, particularly those in a vulnerable situation such as migrants, from data misuses and discrimination (ACLU, 2024). In late 2023, President Biden signed an executive order that established certain guidelines for the proper use of AI and how to prevent privacy risks. However, migrants were only mentioned once for the purpose of "promoting innovation and competition," in order for skilled migrants to be introduced to AI fields (Mengesha et al., 2024).

In Brazil, as part of the Acolhida Operation implemented by the Brazilian government and humanitarian organizations, it has carried out the collection of migrants' personal and biometric data. The objective is to provide identification and assistance to Venezuelans seeking to enter Brazil. Biometric technology was introduced as a solution to diminish the frauds in the humanitarian aid system. However, as Tsui and Shaughnessy (2024) argue, the imposition of this system conveys technological and ethical issues. There are problems using the scanner to capture fingerprints of dark-skinned people and the iris of elderly applicants. In addition, there are ethical concerns considering the application of biometrics to migrants. One of them is the stigmatization of migrants as burdens who want to take advantage of receptive communities. The other is related with the sources that finance the biometric laboratories that capture migrants' data, which are usually Global North countries, as the case of US, which are more interested in controlling and deporting migrants than in providing humanitarian solutions (Corrigan, 2019). Over the years, The US government has signed agreements to obtain the biometric data with all the reception countries where migrants pass during their journey to the USA. This interchange is performed without any transparency of the procedures, and for that reason, some organizations are claiming the end of these practices that compromise the fair applications of migrants during their journey (Access Now, 2023).

Paradoxically, as in the case of the USA, the implementation of biometric captures in Brazilian territory contradicts diverse national regulations and agreements signed by the South American country. Regarding international treaties, Brazil is a part of the Global Compact for Safe, Orderly and Regular Migration (UN, 2018). This deal is based on international human rights law and reaffirms the commitment of states to protect and support the human rights of migrants. The pact's commitments are to guarantee the rights to information and identity of migrants, reducing vulnerabilities and human rights violations in the context of migration, and eliminating discrimination, incitement to hate, and xenophobia. As part of the response to the migration crisis, Brazil, along with other neighboring countries, launched the Brazil Action Plan (2014–2024), a regional roadmap to safeguard human rights of migrants. At a local level, Brazil also counts with proper regulation related with the protection of migrant's rights. The Law No. 13.445 (Presidência da República, 2017). This regulation, also known as the Migration Law, provides a series of safeguards to protect the human rights of migrants. In Section 2, "Principles and Guarantees," Art. 3 refers to the non-criminalization and non-discrimination of migrants in any situation, as well as the compliance with the provisions of international treaties. Article 4, in the same section, guarantees the right to freedom, equality, and security for migrants, along with the guarantee of civil, cultural, and economic rights and freedoms and social security without discrimination based on nationality or migratory status. This article also emphasizes the right to access information and guarantees the confidentiality of the migrant's personal data. These legal stipulations contrast with the extraction of biometric and personal data under the pretext of verifying the migrants' antecedents, and the bias caused by the failure of technologies and the absence of information provided about the use of the data that migrants provide. Regarding the protection of personal data, Brazil established the General Personal Data Protection Law (Presidência da República, 2019) created to protect personal data and to ensure the rights of freedom and privacy, as well as regulating the actions of private and governmental institutions regarding data processing. Recent regulation projects as the case of Bill No. 2338 (Senado

Federal, 2023) establishes the standards for the development, implementation, and responsible use of AI. Art. 2 mentions that the use and development of AI should be based on respect for human rights, equality, non-discrimination, privacy, and data protection. Art. 5 refers to the right to prior information for users regarding their interaction with AI systems and the right to an explanation of the decision or prediction made by AI systems. Regarding the capture of migrants' personal information, the UN Refugee Agency (UNHCR) which has presence across the diverse migration routes to North America, established in 2015 a policy on the Protection of Personal Data of Persons of Interest to UNHCR. This document provides a series of safeguards regarding the collection and management of private data. Section 2.1 for example, stands for a series of principles such as confidentiality, security, accountability, and supervision. Section 3.1 also clarifies that the data subject must be informed if their data will be transferred to third parties. And finally, in section 4.2.1, the UNHCR guarantees the implementation of a high level of security tailored to potential risks that may compromise such information.

Besides counting with a robust national and international legal corpus focused on protecting migrants' human rights, particularly privacy and personal data (Retamozo, 2024), the implementation of biometric captures and the employment of AI systems to manage migrants' applications operates on a kind of "legal limbo" where there is not a designated authority who responds to potential abuses or a data breach. While the operation of these technologies on the ground may be justified to coordinate relief efforts during the migratory path to North America, the lack of transparency on the transfer to third parties and the use of this personal data by migration authorities have generated complaints amid non-governmental organizations dedicated to protecting human rights.

Disruptive Technologies, Migrant Rights, and Civic Mobilization

Across the years, civic organizations that operate at a regional and international level have conducted a series of strategic actions to raise the voice toward the employment of biometric technology, and now artificial intelligence, to manage migration. These activities embrace policy advocacy, litigation, the construction of coalitions, involvement with technological companies that support data interchange, media campaigns to create awareness about the problem and provide evidence of the negative impacts of these technologies in migrant's rights throughout the publication of reports (Prabu & Wisniak, 2024). The following passages detail some of these undertakings.

In 2022, a collective of dozens of civic organizations sent a request to Amazon Web Services (AWS), asking the company to end an agreement with the DHS through which Amazon hosts the database used by the HART (Homeland Advanced Recognition Technology o HART). The organizations, Access Now among them, claimed that the data shared among states occurs with little transparency about the technical and legal protocols to protect user privacy. The prospect of AI to analyze this data is also worrisome due to the potential creation of negative profiles that discriminate against asylum applicants (Alarcón, 2022). AI systems may categorize certain migrants as posing a higher risk and those may have to be further and perhaps unnecessarily investigated. Though the request to Amazon is pending to get a reply, it is a reminder that technological corporations have an important role to play in the protection of human rights. This compromise is even greater when they host the protection of people's rights as a cooperative badge (Okkonen, 2023).

In the promotion of migrants' human rights, the construction of coalitions has been very important for many reasons. One of them is to increase the presence of practitioners across the region since those activists have been able to gather information and create consciousness about the problem at a local and international level. The coalitions also help to widen the voice in international forums and across the Internet, and to submit official requests of information about technical procedures to international organizations, technology companies, and governments (R3D and Access Now, 2023). In the conformation of alliances and the process of creating awareness, the employment of social media has been a key issue, this not only because it enhances the reach of the organization's agenda, but also facilitates the conformation of greater social movements. By the end of 2024 for example, Centro LATAM Digital (2024), with the aim to publicize the launch of a report, set the campaign socially, #MigrarSinVigilancia. In this case, the use of hashtags implies not only a strategy to make viral the message, but also to incorporate most organizations onto the movement. However, as Prabu & Wisniak (2024) argue is that the most effective campaign toward biometric surveillance appears when diverse tactics are undertaken at the same time as collecting evidence and spreading the information through diverse information channels such as social media, brief frameworks and wider publications as the case of reports. Equally, Internet awareness actions could be undertaken along with formal litigations at human rights courts (Prabu & Wisniak, 2024).

For years, migration authorities in the US have used AI, ignoring legal regulations and ethical limits regarding private abuses (Hu, 2022), however this is starting to change. The concerns and criticism about biometric captures and AI facial recognition technology settled by non-governmental organizations have reached the US senate. On November 2024, a bipartisan group sent a document to the DHS expressing their apprehensions about the extensive use of facial recognition technology at US airports without guarantees about the protection of privacy (Smalley, 2024). This testimony remarks well how AI technology, besides their benefits, should be applied under legal and technical protocols to protect people's rights. This is particularly important considering people who are in compromised situations such as refugees. But it is also important to understand that disruptive technologies, even with the intention to increase homeland security toward international threats such as terrorism, could not be a free pass to using it without well-established protocols. In addition, AI-collected data from migrants, although taken from governmental agencies in which the people have the notion of trust and has been well-intentionally collected, may be compromised by third-party cyberattacks. Although AI can be beneficial to speed up application processing and shorter security-screening wait times, it could also promote discrimination, bias, and further worsen existing inequalities. The potential denial of migrant applications due to a faulty AI screening can cause further frustrations to an already overwhelmed migrant. Seeking clarification on a potential AI error may be vague or there may not be clear instructions on how to proceed. Filing a complaint may have several obstacles in addition to being a lengthy and time-consuming process, and the migrant might not have the capacity or the tools to further continue. Re-filing a wrongfully denied or discriminated against application may not be possible or may pose additional challenges for the migrants.

Conclusions

The cases analyzed in these chapters testified to the urgency to focus on the legal and ethical problems caused by the increasing use of AI to manage data, particularly when these systems are taking a decisive role in the decision-making process about migration. While the United States, Brazil and the international organizations that participate in the capturing of migrants' information departs from the safeguard of people as the main objective by introducing these technologies, in practice the procedure has been used to criminalize and deny the entry to the United States using unclear technological presets (Nalbandian, 2022). The academic literature and punctual cases studied in this work confirm that the US authorities are using that data gathered on applications such as CBP One and are obtained and shared by Latin American countries to exclude migrants from their rights to seek asylum, not as an instrument to protect migrants during their journey. The case of Brazil illustrates well how the capturing of people's information, as the case of Venezuelans, occurs in situations where the migrants are not well informed about the destination of their personal data, or the existence of essential legal and technical safeguards that allow to delete their data from the systems or how this information could be used for a future rejection of their application (Amnesty International, 2024).

Another intriguing finding of this research focuses on how besides the existence of a robust legal corpus at a national and international level related with the protection of human rights, particularly privacy, and even technological firms such as Amazon embrace these principles as a cooperative insignia. The capturing, transferal and usage of migrants' data occurs under uncertain procedures that do not guarantee the integrity of the information, and many times employed to practices not authorized by the owner (Bircan & Korkmaz, 2021). As observed in this work, during these years, the AI and biometric technologies established in the migration control have operated, omitting national and international regulations on the subject and compromising petitioners' rights. Another important asset, as it is possible to observe in other volume chapters, is the absence of specialized regulation on the subject. As it has happened with the incorporation of other devices and applications based on AI such as drones, AI cameras or ChatGPT, governmental agencies have incorporated them in their activities without counting with a proper regulation on the subjects and the establishment of technological guidelines to operate AI (Holloway et al., 2021; Alarcón et al., 2024). This is problematic when sensitive government and citizen information is introduced as raw data in this AI software or there are not enough technological protocols to protect the integrity of material captured by AI devices. This work agrees with previous research that claims that the implementation of technology to improve migration systems should be focused not only on the technological aspect, but also, as many national policies and international frames have declared, centered on human thinking and based on the protection of user's rights (Gottardo et al., 2024; Roveri, 2024).

References

Access Now. (2023, March 23). Sociedad civil exige la terminación de los acuerdos para el tratamiento transfronterizo de datos biométricos de personas migrantes. *Access Now*. Retrieved January 6, 2025, from https://www.accessnow.org/press-release/sociedad-civil-exige-terminar-acuerdos-biometricos-migrantes/

ACLU. (2024, October 24). ACLU Warns that Biden-Harris Administration Rules on AI in National Security Lack Key Protections. *American Civil Liberties Union*. Retrieved January 9, 2025, from https://www.aclu.org/press-releases/aclu-warns-that-biden-harris-administration-rules-on-ai-in-national-security-lack-key-protections

ACNUR. (2021, May 3). Brasil: Operación Acogida. *Buenas Practicas Asilo Americas*. Retrieved January 8, 2025, from https://www.asiloamericas.org/bra-operacion-acogida/

Alarcón, Á. (2022, September 21). Exigimos que las personas puedan #MigrarSinVigilancia. *Access Now*. Retrieved January 6, 2025, from https://www.accessnow.org/exigimos-que-las-personas-puedan-migrar-sin-vigilancia/

Alarcón, Á., Giandana Gigena, F., & Coppi, G. (2024, March 13). "Smart borders" and the making of a humanitarian crisis. *Access Now*. Retrieved January 9, 2025, from https://www.accessnow.org/smart-borders-and-the-making-of-a-humanitarian-crisis/

Amnesty International. (2024). *Primer: Defending the Rights of Refugees and Migrants in the Digital Age*. Amnesty International. Retrieved January 6, 2025, from https://journals.sagepub.com/doi/10.1177/2056305118764432

Avraamidou, M., & Eftychiou, E. (2022). Migrant Racialization on Twitter during a Border and a Pandemic Crisis. *International Communication Gazette, 84*(3), 227–251. https://doi.org/10.1177/17480485211054301

Bhuiyan, J. (2023, May 25). Revealed: The contentious tool US immigration uses to get your data from tech firms. *The Guardian*. Retrieved January 10, 2025, from https://www.theguardian.com/us-news/2023/may/25/us-immigration-surveillance-google-twitter-meta-personal-data

Biden, J. (2024). *Memorandum on Advancing the United States' Leadership in Artificial Intelligence; Harnessing Artificial Intelligence to Fulfill National Security Objectives; and Fostering the Safety, Security, and Trustworthiness of Artificial Intelligence*. Retrieved January 9, 2025, from https://www.whitehouse.gov/briefing-room/presidential-actions/2024/10/24/memorandum-on-advancing-the-united-states-leadership-in-artificial-intelligence-harnessing-artificial-intelligence-to-fulfill-national-security-objectives-and-fostering-the-safety-security/

Bircan, T., & Korkmaz, E. E. (2021). Big Data for Whose Sake? Governing Migration through Artificial Intelligence. *Humanities and Social Sciences Communications, 8*(1), 1–5. https://doi.org/10.1057/s41599-021-00910-x

Camargo, J., & Alencar, A. (2024, May). Perspectivas de la población venezolana sobre las tecnologías fronterizas. *Forced Migration Review*. Retrieved January 8, 2025, from https://www.fmreview.org/disrupcion-digital/camargo-alencar/

Centro LATAM Digital. (2024, December 17). #MigrarSinVigilancia: Un Llamado Urgente para Proteger los Datos de las Personas Migrantes. *LATAM DIGITAL*. Retrieved January 6, 2025, from https://centrolatam.digital/migrarsinvigilancia-un-llamado-urgente-para-proteger-los-datos-de-las-personas-migrantes/

Congress.gov. (2025). U.S. Constitution—Fourth Amendment. *Constitution Annotated*. Library of Congress. Retrieved January 9, 2025, from https://constitution.congress.gov/constitution/amendment-4/

Corrigan, J. (2019, August 20). DHS is Collecting Biometrics on Thousands of Refugees Who Will Never Enter the U.S. *Nextgov.com*. Retrieved January 9, 2025, from https://www.nextgov.com/emerging-tech/2019/08/dhs-collecting-biometrics-thousands-refugees-who-will-never-enter-us/159310/

Dávila, S., Palacios, A., & Thompson, B. (2024). *CBP One Mobile Application: Violating Migrants' Rights to Privacy CBP One Mobile Application: Violating Migrants' Rights to Privacy and Freedom from Discrimination and Freedom from Discrimination*. UIC Law White Papers.

Dawes, J. (2021, September 29). An autonomous robot may have already killed people—Here's how the weapons could be more destabilizing than nukes. *The Conversation*. Retrieved December 25, 2024, from http://theconversation.com/an-autonomous-robot-may-have-already-killed-people-heres-how-the-weapons-could-be-more-destabilizing-than-nukes-168049

Dib, D., & Deslandes, A. L. (2023, March 9). U.S. Border Patrol's glitchy, mandatory app is a nightmare for migrants. *Rest of World*. Retrieved January 9, 2025, from https://restofworld.org/2023/cbp-one-app-issues-migrants/

Felbab-Brown, V. (2021, September 10). 9-11 and the US-Mexico border: New challenges 20 years later. *Brookings*. Retrieved January 8, 2025, from https://www.brookings.edu/articles/9-11-and-the-us-mexico-border-new-challenges-20-years-later/

France 24. (2023, June 28). Ciencia y Tecnología—Vigilancia biométrica, una práctica que violaría los derechos de los migrantes. *France 24*. Retrieved December 25, 2024, from https://www.france24.com/es/programas/ciencia-y-tecnolog%C3%ADa/20230628-vigilancia-biom%C3%A9trica-una-pr%C3%A1ctica-que-violar%C3%ADa-los-derechos-de-los-migrantes

France 24. (2024, June 4). ¿En qué consiste la drástica medida migratoria anunciada por Biden? *France 24*. Retrieved July 30, 2024, from https://www.france24.com/es/ee-uu-y-canad%C3%A1/20240604-biden-firma-su-medida-migratoria-m%C3%A1s-dura-y-limitar%C3%A1-las-solicitudes-de-asile-en-la-frontera-con-m%C3%A9xico

Gottardo, C., Finch, C., & Cooper, H. (2024, June 25). Digital Technology, Detention and Alternatives—International Detention Coalition. In *Digital Technology, Detention and Alternatives*. IDC. Retrieved January 8, 2025, from https://idcoalition.org/digital-technology-detention-and-alternatives/

Holloway, K., Masri, R. A., & Yahia, A. A. (2021). *Digital Identity, Biometrics and Inclusion in Humanitarian Responses to Refugee Crises* (Working Papers). ODI: Think Change. Retrieved January 6, 2025, from https://odi.org/en/publications/digital-identity-biometrics-and-inclusion-in-humanitarian-responses-to-refugee-crises/

Hu, M. (2022). Biometrics and an AI Bill of Rights. *Duquesne Law Review, 60*, 283.

Kinchin, N. (2024). The Risk of Risk: Algorithms and Immigration Detention. *UNSW Law Society Court of Conscience, 18*, 129–135.

Koch, A. (2023). The International Organization for Migration as a Data Entrepreneur: The Displacement Tracking Matrix and Data Responsibility Deficits. In A. Sherwood, C. Costello, & M. Bradley (Eds.), *IOM Unbound?: Obligations and Accountability of the International Organization for Migration in an Era of Expansion* (pp. 235–269). Cambridge University Press. https://doi.org/10.1017/9781009184175.012

Korkmaz, E. E. (Ed.). (2021). *Digital Identity, Virtual Borders and Social Media: A Panacea for Migration Governance?* Edward Elgar Publishing. https://doi.org/10.4337/9781789909159

Lemos, A. (2021, April 15). Brazil Installs Intelligent Border Control Systems. *Diálogo Américas*. Retrieved January 8, 2025, from https://dialogo-americas.com/articles/brazil-installs-intelligent-border-control-systems/

Nalbandian, L. (2022). An Eye for an 'I:' A Critical Assessment of Artificial Intelligence Tools in Migration and Asylum Management. *Comparative Migration Studies, 10*(1), 32. https://doi.org/10.1186/s40878-022-00305-0

Nedelcu, M., & Soysüren, I. (2022). Precarious Migrants, Migration Regimes and Digital Technologies: The Empowerment-Control Nexus. *Journal of Ethnic and Migration Studies, 48*(8), 1821–1837. https://doi.org/10.1080/1369183X.2020.1796263

Okkonen, L. (2023, May 18). Amazon shareholders. *Access Now*. Retrieved January 10, 2025, from https://www.accessnow.org/amazonshareholders/

Prabu, A., & Wisniak, M. (2024). *Taking Action Against Biometric Surveillance Civil Society Tactics and Strategies*. The European Center for Not-for-Profit Law (ECNL). Retrieved January 6, 2025, from https://ecnl.org/sites/default/files/2024-11/ECNL%20Biometrics%20Surveillance%20Strategies%202024_v%2030.09.24_0.pdf

Presidência da República. (2017). *L13445*. Retrieved January 9, 2025, from https://www.planalto.gov.br/ccivil_03/_ato2015-2018/2017/lei/l13445.htm

Presidência da República. (2019). *Lei Geral de Proteção de Dados Pessoais (LGPD)*. Retrieved January 9, 2025, from https://www.planalto.gov.br/ccivil_03/_ato2015-2018/2018/lei/l13709.htm

R3D & Access Now. (2023). *Civil society call: Terminate agreements on migrants' biometric data*. Access Now/la Red en Defensa de los Derechos Digitales (R3D). Retrieved January 6, 2025, from https://www.accessnow.org/press-release/civil-society-demands-termination-of-agreements-on-migrants-biometric-data/

Ramírez Plascencia, D. (2024). Unwelcomed Neighbors: Media Coverage and the Spread of Hate Towards Migrants in Latin America. *REMHU: Revista Interdisciplinar da Mobilidade Humana, 32*, 1–16. https://doi.org/10.1590/1980-85852503880003215

Retamozo, B. I. M. (2024). Border Surveillance: Biometric Technologies and Personal Data in Latin American Migration. *Epistemia Revista Científica, 8*(2), 1–16. https://doi.org/10.26495/erc.2820

Roveri, C. (2024, March 15). Derechos las personas refugiadas y migrantes en la era digital. *Amnistía Internacional*. Retrieved January 6, 2025, from https://www.es.amnesty.org/en-que-estamos/reportajes/derechos-personas-refugiadas-y-migrantes-en-la-era-digital/

Saunders, N. (2023). Security, Digital Border Technologies, and Immigration Admissions: Challenges of and to Non-discrimination, Liberty and Equality. *European Journal of Political Theory, 24*, 14748851231203912. https://doi.org/10.1177/14748851231203912

Senado Federal. (2023). *Projeto de Lei N° 2338*. Retrieved from https://legis.senado.leg.br/sdleg-getter/documento?dm=9347622&ts=1733409396734&disposition=inline

Smalley, S. (2024, November). Senators call for audit of TSA's facial recognition tech as use expands in airports. *The Record*. Retrieved January 6, 2025, from https://therecord.media/tsa-facial-recognition-tech-senators-call-for-audits

Mengesha, S., Dunn, K., & Luangrath, N. (2024, March 23). The Rise of AI and Technology in Immigration Enforcement. *The Regulatory Review*. Retrieved from https://www.theregreview.org/2024/03/23/the-rise-of-ai-and-technology-in-immigration-enforcement/

Tsui, Q., & Shaughnessy, E. (2024, May). Tecnocolonialismo y biometría: Un nuevo impulso al llamamiento para descolonizar la ayuda. *Forced Migration Review*. Retrieved January 9, 2025, from https://www.fmreview.org/disrupcion-digital/tsui-shaughnessy/

Tucker, P. (2024, December 2). New AI tool for air defense takes on advanced missiles and drone swarms. *Defense One*. Retrieved December 25, 2024, from https://www.defenseone.com/technology/2024/12/new-ai-tool-air-defense-takes-advanced-missiles-and-drone-swarms/401381/

U.S. Customs and Border Protection. (2023, August 3). Artificial Intelligence to Harness Key Insights at CBP. *Artificial Intelligence to Harness Key Insights at CBP.* CBP. Retrieved January 8, 2025, from https://www.cbp.gov/newsroom/spotlights/artificial-intelligence-harness-key-insights-cbp

UN. (1965). *International Convention on the Elimination of All Forms of Racial Discrimination.* Retrieved January 9, 2025, from https://www.ohchr.org/en/instruments-mechanisms/instruments/international-convention-elimination-all-forms-racial

UN. (1976). *International Covenant on Civil and Political Rights.* Retrieved January 9, 2025, from https://www.ohchr.org/sites/default/files/ccpr.pdf

UN. (2018). *Global Compact for Safe, Orderly and Regular Migration.* Retrieved January 9, 2025, from https://www.iom.int/global-compact-migration

US Congress. (1988). *Computer Matching and Privacy Protection Act.* Retrieved January 9, 2025, from https://www.congress.gov/bill/100th-congress/senate-bill/496

Ethical and Educational Dilemmas of AI in Latin American Higher Education Institutions: Persistent Challenges and Inquiries

Jairo Alberto Galindo-Cuesta, Rubén Yáñez Reyna, and Paola Mercado Lozano

Introduction

The humanoid robot Ai-Da, renowned for her groundbreaking artistry, recently unveiled a thought-provoking piece titled "A.I. God." (Sotheby's, 2024). This work, centered on Alan Turing, the father of modern computing, invites audiences to reflect on the divine and transformative nature

J. A. Galindo-Cuesta (✉)
National Distance Education University (UNED), Madrid, Spain
La Salle University, Bogotá, Colombia
e-mail: jairogalindo@unisalle.edu.co

R. Yáñez Reyna • P. Mercado Lozano
University of Guadalajara, Guadalajara, Jalisco, Mexico
e-mail: ruben.yanez@academicos.udg.mx; paola.mercado@cugdl.udg.mx

D. Ramírez Plascencia, R. M. Alonzo González (eds.), *Ethical and Legal Dilemmas of Artificial Intelligence in Latin America*,
https://doi.org/10.1007/978-3-031-86540-4_8

of artificial intelligence. By addressing Turing's legacy and its ethical and social implications, Ai-Da bridges the realms of technology and humanity, prompting discussions about the role of advanced AI in shaping our societal values and moral frameworks. Such reflections resonate deeply within educational contexts, particularly in Latin America, where higher education institutions face persistent ethical and pedagogical dilemmas concerning the integration of generative AI. This chapter delves into the challenges and inquiries surrounding the implementation of ethical guidelines for AI in education, offering insights into how these frameworks can address emerging complexities in university settings.

In contemporary educational settings, questions persist about how well institutions can meet the diverse needs of all students, making it challenging to determine the most suitable content and strategies for different situations (Levinson & Fay, 2016; Tubella et al., 2023). Tesar and his team (2021) highlight that "interconnected ecological systems of ethics have not emerged through a single rational process," underscoring the fragmented nature of ethical frameworks in AI. In various university settings, documents are being created (Jiang & Zhou, 2024) to minimize the confusion caused by the haphazard use of artificial intelligence methods. In this context, UNESCO advocates for establishing rules and guidelines to ensure that generative artificial intelligence (GenAI) is a tool that genuinely enhances the capabilities of educators, learners, and researchers. While there are studies on related topics (McMahon & Milligan, 2021; Dabis & Csáki, 2024), this work examines how ethical guidelines are implemented in some Latin American universities (Brazil, Mexico, and Colombia) when using artificial intelligence in education and research. The aim is to identify how the formulation of a cross-cutting ethical framework in education may share common elements with the discerning of the potentialities and challenges posed by GenAI in didactics and pedagogy.

Theoretical Framework

The Role of Ethics in GenAI

The ethical implementation of GenAI (Generative AI) in education is grounded in various technological and pedagogical disciplines that address the challenges and opportunities caused by the incorporation of this so-called disruptive technology. From a philosophical perspective, Floridi (2019) explores how the principles of information and conceptual logic can serve as a basis for understanding and addressing the ethical implications of GenAI. This approach is particularly relevant in education, where values such as justice, equity, and dignity must guide the implementation of disruptive technologies. Jobin et al. (2019) highlight the importance of transparency and data protection as fundamental pillars to ensure trust in technological systems. Recently, algorithmic bias and its impact on educational equity have been the subject of significant studies. Binns (2018) analyzes how lessons from political philosophy can support the design of fairer algorithmic systems, while O'Neil (2016) demonstrates how automated decisions can perpetuate inequalities if not implemented with clear ethical criteria. This debate is crucial in education, where equitable access to GenAI can either enhance or limit learning opportunities.

To face the analysis and discussion on the ethical dimension of GenAI in education, it is necessary to briefly define it. GenAI is a specific subfield within artificial intelligence, which, according to Turing (1950), is understood as the ability of machines to mimic human intelligence (Copeland, 2024). GenAI employs advanced language models capable of generating text, images, voices, code, and more, in response to user requests expressed in natural language (Gallent Torres et al., 2023). Andión Gamboa and Cárdenas Presa (2023) categorize GenAI into three types based on its function "consultant tools, virtual assistants (copilots), and agents (robots) capable of performing tasks automatically and making decisions based on algorithms" (p. 58). This approach grants GenAI a creative and adaptive capacity that distinguishes it from other applications, making it particularly relevant for educational contexts and the ethical debate surrounding its use.

The ethical debate around GenAI in education is closely linked to notions such as equity, integrity, and transparency. Tauginiené et al. (2018) define academic integrity as the adherence to ethical and professional principles that, together with a consistent system of values, guide decision-making in education, research, and study (p. 8). This concept is grounded in essential values such as honesty, responsibility, transparency, and trust (Rodríguez et al., 2024). These principles, which are central to education, must be respected and guaranteed by the entire educational community, regardless of the teaching modality adopted, including the use of emerging technologies like GenAI.

Authors such as Currie (2023) and Rodríguez et al. (2024) argue that academic integrity faces significant challenges in the era of AI, as it presents both opportunities and limitations. Among the main limitations are the risks for both educators and students, such as academic fraud and plagiarism. Academic dishonesty can manifest in various ways, including plagiarism, inappropriate collaboration, cheating in exams, violations of copyright, and data fabrication, among others. Regarding equity, Albuquerque et al. (2023) argue that unequal access to technology in education exacerbates social and economic disparities and increases marginalization among historically excluded populations (p. 98). Similarly, the OECD (Kuhl et al., 2019) emphasizes that the lack of access to technology can create a digital divide that deepens educational inequality.

The Reformulation of Theoretical Definitions

The rapid evolution of artificial intelligence (AI) has led to the redefinition of key theoretical frameworks, particularly with the introduction of Generative AI, a technology capable of creating new content from existing data. These advancements have reshaped how concepts such as technological sovereignty and inclusive AI are understood and applied in regions like Latin America.

Technological Sovereignty has become a central concept, reflecting the need for nations to reduce dependency on foreign technologies and foster local innovation. Floridi (2019) emphasizes that technological sovereignty is not only a matter of innovation but also a necessity for ethical autonomy in AI development. Brazil exemplifies this approach with its historic investment of 23 billion reais (approximately 4 billion USD) announced in July 2024 (Reuters, 2024). This initiative underscores the country's commitment to achieving AI independence, aligning with broader efforts to position Brazil as a leader in AI innovation while securing its technological future.

Inclusive and Contextualized AI highlights the importance of tailoring AI systems to the socio-economic and cultural realities of their users. Ożegalska-Łukasik and Łukasik (2023) stress that inclusive AI must address structural inequalities, particularly in regions marked by significant disparities. In Mexico, efforts to democratize AI within its education system reflect this principle, aiming to equip future generations with the skills needed to interact with advanced technologies (SIPSE, 2024). These initiatives illustrate the region's focus on bridging the digital divide and fostering equitable access to AI tools.

Legislation on AI in Europe and Latin America

The need to govern AI technologies, including generative AI (GenAI), has prompted legislative efforts worldwide. The European Union has been at the forefront of these initiatives, establishing foundational frameworks to ensure the ethical and secure development of AI systems. Jobin et al. (2019) argue that transparency and accountability are critical pillars in building trust in AI technologies, a perspective reflected in the EU's legislative progress. In 2020, the EU introduced a civil liability framework for AI, which culminated in the 2024 publication of the AI Act (European Union, 2024). This comprehensive regulation emphasizes ethical principles, security, and the protection of fundamental rights, setting a global benchmark for AI governance.

Latin America, while still in the early stages of regulatory development, is making strides to align with these international standards. Brazil stands out as a pioneer, even ahead of the European Union. In 2019, the first legislative proposals were launched in the National Congress, and in 2024, the Bill (PL) No. 2.338/2023 is being promoted for approval by the end of the year (Albuquerque et al., 2023). Colombia developed the Ethical Framework for AI in 2020 (Arce González et al., 2023), while in Mexico, regulation in AI matters is still in process in 2024, through debates between civil society, government, and international organizations (Medrano Rodríguez et al., 2024).

Considering higher education, international organizations have joined the initiative to regulate and establish ethical standards for the use of GenAI in education. On May 18, 2019, UNESCO published the Beijing Consensus on AI and Education, the first document that provides recommendations on the use of AI technologies (UNESCO, 2019). Two years later, on November 23, 2021, UNESCO approved the Recommendation on the Ethics of Artificial Intelligence, the first global norm on AI ethics, adopted by 193 member states (UNESCO, 2021). This document examines the ethical implications of AI use in various areas of activity (Gallent Torres et al., 2023). Regarding educational institutions, some universities have established recommendations on the use of Generative Artificial Intelligence (GenAI) in teaching and research.

The National Autonomous University of Mexico (UNAM, 2023) issued a document that transitions from experimentation to the critical appropriation of technology. It highlights security challenges in implementing GenAI in teaching and academic ethics. Furthermore, it emphasizes the necessity of integrating technology creatively into learning spaces and processes, including assessment. Finally, it proposes that, in an initial phase, GenAI be utilized for automating routine teaching activities (UNAM, 2023). The National Polytechnic Institute has undertaken the task of designing two postgraduate programs: a master's and a doctorate in Artificial Intelligence Science and Technology and Data Science, which will be offered in 2025. The institution has also published works on GenAI applications in educational processes, such as language learning (2024) and others (IPN, 2021). The Autonomous University of Puebla organized a series of conferences in 2024 to explore GenAI and its challenges in the publishing industry and education. The objective was to gather diverse perspectives and tools to confront the challenges posed by this technology (BUAP, 2024). The Autonomous University of San Luis Potosí established a group in 2023 to analyze the challenges and opportunities of AI in education. The goal was to provide the university community with information on best practices and recommendations for incorporating GenAI into educational processes (UASLP, 2023) The Metropolitan Autonomous University hosted an academic forum in 2024 focused on the multidisciplinary perspectives of AI in higher education (UAM, 2024). Lastly, the University of Guadalajara published a practical guide in 2023 containing guidelines and definitions on the use of GenAI in academic processes. This guide, aimed at educators at all levels, serves as a starting point for understanding general information about GenAI, its ethical challenges, and potential applications (UDG, 2023). In Brazil, diverse universities are already joining efforts to establish ethical frameworks to employ generative AI in academic research (Schmidt, 2024). In recent years, Colombian universities such as UNIADES, Universidad del Norte, and Universidad Externado de Colombia have also published documentations and guidelines to orient students and processors toward an efficient and ethical employment of AI in the classroom.

GenAI in the Context of Higher Education in Latin America

To comprehend the status of GenAI in Latin America, particularly in the countries selected for this study (Brazil, Colombia, and Mexico), it is relevant to analyze the data provided by the Latin American Artificial Intelligence Index (LAIA) 2024. This index offers quantitative and qualitative information on the state of AI development in 19 Latin American and Caribbean countries, allowing the identification of achievements, gaps, and opportunities for improvement in various dimensions related to AI (LAIA, 2024).

Considering the countries included in this study (Brazil, Colombia, and Mexico), Brazil ranks second in the LAIA (2024), placing it in the "pioneer" category. This category represents a high level of maturity in key dimensions such as technological infrastructure, talent development, scientific productivity, and innovation capacity. Being considered a "pioneer" implies that Brazil has achieved a leadership position in AI in the region. Colombia and Mexico rank fifth and sixth, respectively, placing them in the "adopters" category. This level reflects significant progress in integrating AI into various sectors, although they have not yet reached the maturity or leadership of pioneer countries. In the education sector, Brazil leads the region in the Research, Adoption, and Development dimensions included in the LAIA (2024), occupying the first position. Mexico ranks fourth, followed by Colombia in sixth place. These results help to understand the strengths and areas for improvement in each country regarding education, research, and innovation, with Brazil serving as a regional reference, and Mexico and Colombia showing significant progress (LAIA, 2024).

In the Latin American context, discussions on AI ethics and its educational impact reflect the sociocultural and economic particularities of the region. Botero and Sáenz (2022) highlight how the digital divide is amplified with the use of advanced technologies if not addressed from a perspective of social justice. Complementary to this, Cobo (2016) emphasizes the need to foster critical and creative skills in students to ensure ethical and meaningful use of technology.

Studies in the region also highlight a growing concern about data protection and privacy in the educational context. UNESCO (2023) underscores that educational institutions must adopt robust regulatory frameworks to safeguard personal information of students and teachers while promoting responsible and transparent use of AI. Blackwell (2024) provides a critical perspective, highlighting how power dynamics and corporate interests can influence the implementation of educational technologies, requiring an ethical and contextualized approach. Ramírez and Casillas (2024) conducted a study to understand the perceptions and experiences of Mexican teachers regarding the use of GenAI in education.

By surveying 3500 participants, they found that the main concerns expressed by teachers are plagiarism, laziness, and technological dependency. In summary, these theoretical references and empirical studies contribute to a comprehensive understanding of the ethical challenges associated with GenAI in education. They also provide a solid foundation for formulating recommendations that ensure responsible, inclusive, and aligned adoption with the principles of holistic human development.

Methodology and Data Collection

The study[1] employed a mixed-methods approach, combining qualitative and quantitative data collection and analysis. The research was conducted in collaboration with five universities across Latin America, selected to represent a range of institutional types and regional contexts. Sample. A total of 138 faculty members were surveyed to gather data on their perceptions of AI use in education and research, as well as their awareness of and adherence to ethical guidelines.

Data Analysis

The data were analyzed using thematic analysis to identify key themes and patterns. Quantitative data were analyzed using descriptive and inferential statistics to assess trends and correlations.

Participant Profile

The survey was sent via email to faculty members at private higher education institutions in Mexico, Colombia, and Brazil. A total of 212 participants responded, but only 138 completed the questionnaire in its entirety. Regarding gender, 66 teachers were female, and 71 were male. In terms of educational level, most teachers held a master's degree (70), followed by 42 with a Doctorate and 8 with Postdoctoral studies.

[1] This section presents some outcomes of a major research conducted by the authors in collaboration with the RIDUL Group at La Salle Universities.

The results presented below reflect a detailed analysis of the ethical perceptions and practices related to the use of GenAI in education. Through an instrument applied to a representative group of teachers, the study aimed to identify the main ethical concerns and levels of action in key areas such as privacy, transparency, equity, and academic integrity. This analysis not only allows understanding the predominant attitudes toward ethical challenges but also reveals existing gaps in terms of training, institutional policies, and awareness of the impact of GenAI on pedagogical dynamics. Additionally, the results highlight both the progress achieved in some areas and the persistent challenges, providing empirical evidence to propose improvement strategies for the ethical integration of these technologies in educational contexts.

Outcomes

To facilitate a thematic analysis, the questions were grouped into three categories, each representing interconnected ethical dimensions. The following results reflect the aggregated data:

Group 1: Privacy, Equity, and Ethical Actions (Fig. 1).

- Have you taken any action to address ethical concerns or responsibilities related to GenAI? Yes (32.6%), No (22.5%), N/A (43.5%).
- Do you consider privacy and data protection norms ethical challenges? Yes (73.9%), No (26.1%).
- Do you consider equity and non-discrimination ethical challenges? Yes (27.5%), No (71.7%).

Group 2: Transparency, Consent, and Integrity (Fig. 2).

- Do you consider transparency and accountability ethical challenges? Yes (45.7%), No (54.3%).
- Do you consider informed consent an ethical challenge? Yes (51.4%), No (47.8%).
- Do you consider academic integrity an ethical challenge? Yes (81.2%), No (18.8%).

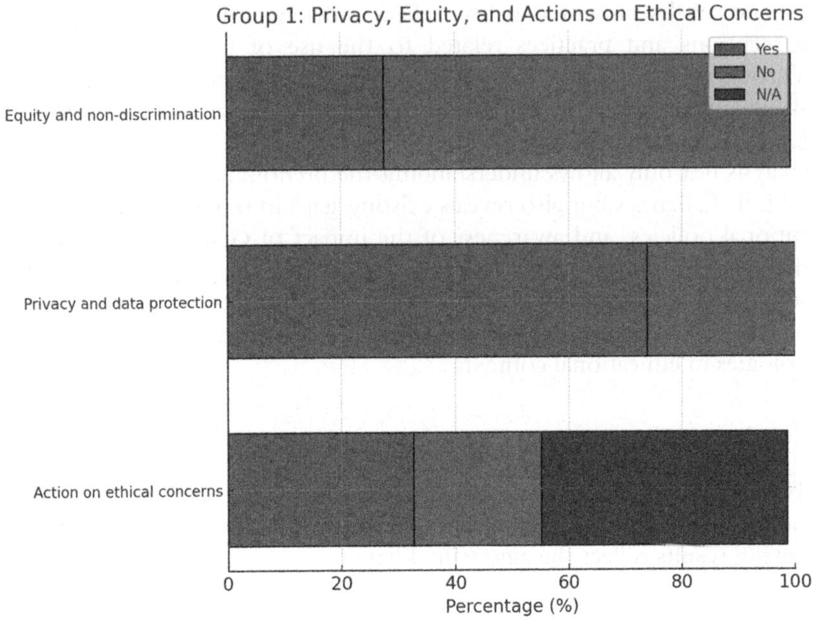

Fig. 1 Group 1. Privacy, equity, and ethical actions

Group 3: Access, Development, and Other Concerns (Fig. 3).

- Do you consider equitable access to AI technologies an ethical challenge? Yes (19.6%), No (79.7%).
- Do you consider continuous professional development an ethical challenge? Yes (52.2%), No (47.1%).
- Do you consider environmental impact an ethical challenge? Yes (9.4%), No (89.9%).
- Do you consider other ethical challenges associated with AI in education? Yes (0%), No (1.4%).
- Do you implement any guidelines for the use of GenAI in your teaching practices? Yes (30.4%), No (68.1%).

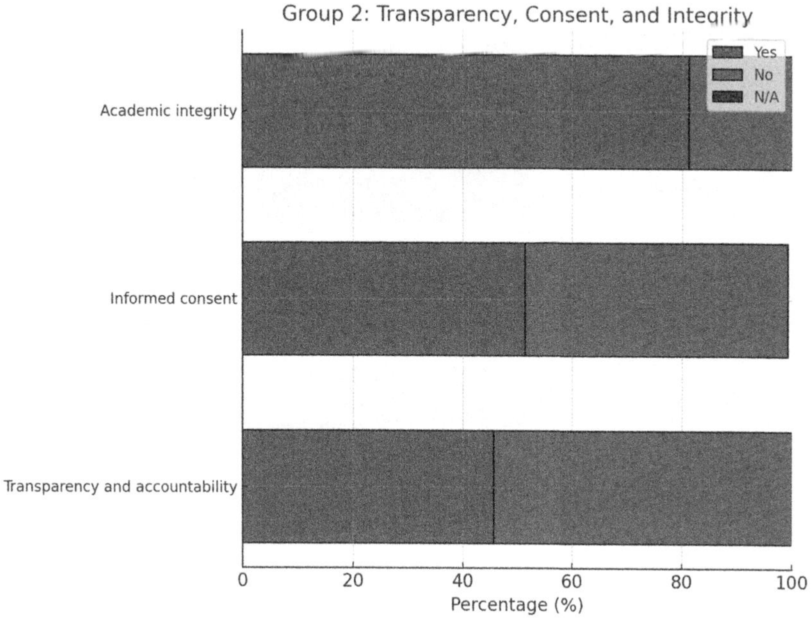

Fig. 2 Group 2. Transparency, consent, and integrity

Discussion

The study's outcomes underscore the intricate and multifaceted nature of implementing ethical frameworks for generative artificial intelligence (GenAI) within higher education institutions across Latin America. One of the most significant observations is the disparity in understanding and the application of ethical principles among faculty members in the surveyed universities. While there is widespread recognition of the importance of addressing issues such as academic integrity, equity, and privacy, the extent to which these principles are translated into actionable guidelines remains inconsistent. This divergence highlights a critical gap between theoretical awareness and practical implementation.

None of the institutions participating in the study has prepared and implemented any (official) policy by the date of this research. Except for proposals for courses, training or brief talks, none of them had incorporated the ethics of artificial intelligence in their discourses, which also

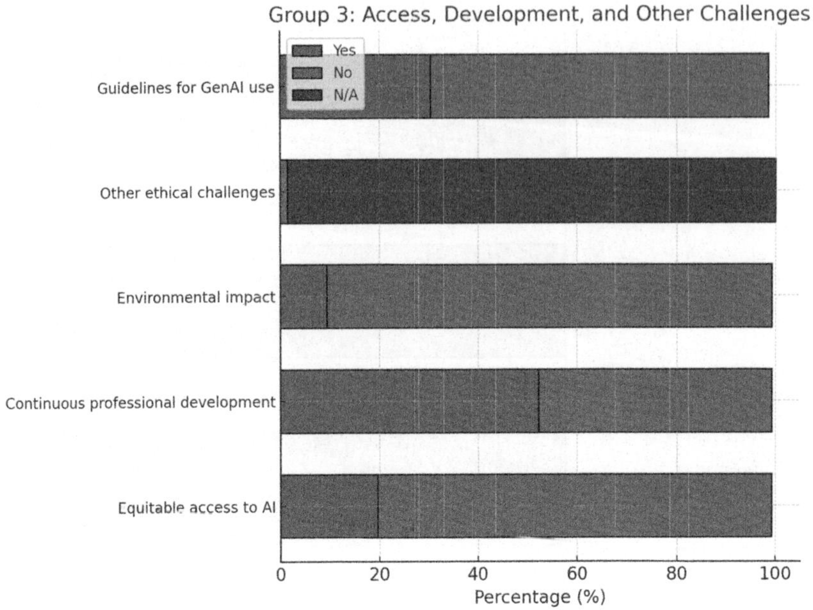

Fig. 3 Group 3. Access, development, and other concerns

allows highlighting the findings and antecedents cited in sharing how ethics can be in the reflection, but not in the practice or in the official institutional discourse.

The role of institutional culture emerged as a decisive factor in shaping attitudes toward GenAI. Institutions with a stronger emphasis on ethical leadership and comprehensive professional development programs were more likely to exhibit proactive measures in incorporating GenAI responsibly. However, even in these cases, challenges such as insufficient infrastructure, limited funding, and a lack of regionally tailored ethical policies persist. These issues are particularly salient in countries like Colombia and Mexico, which, despite significant strides in AI adoption, often lack the systemic support to align technological advancements with ethical imperatives.

Furthermore, the concept of digital equity remains an urgent concern. The uneven access to AI technologies, exacerbated by socio-economic inequalities, creates a dichotomy between those who can harness the benefits of GenAI and those who remain marginalized. This digital divide not only limits the potential impact of GenAI on education but also risks entrenching existing disparities. The findings resonate with prior studies (Albuquerque et al., 2023; UNESCO, 2021) that emphasize the necessity of public policies aimed at democratizing access to digital tools and fostering digital literacy among historically underserved populations.

A particularly intriguing insight is the tension between innovation and regulation; there are institutions that are only beginning to engage in discussions with external stakeholders to make decisions regarding the incorporation of Generative AI (GenAI) and its corresponding regulatory framework. In contrast, other institutions have worked internally with their experts to design proposals that facilitate the adoption of GenAI in teaching at various educational levels. While universities and policymakers alike recognize the transformative potential of GenAI, there is a palpable hesitancy to embrace it fully due to concerns over its ethical implications. This caution is compounded by the absence of harmonized regulatory frameworks at both national and regional levels. The divergence in legislative progress among Brazil, Colombia, and Mexico exemplifies the challenges of achieving a cohesive approach to AI governance in the region. Brazil's pioneering efforts, including its legislative proposals and investment in AI infrastructure, contrast sharply with the nascent and fragmented initiatives in other countries.

It is important to consider the establishment of standards as an option to create conditions and guidelines for the implementation and use of AI systems. ISO standards, internationally recognized for defining quality, safety, and sustainability criteria, can provide elements for unifying criteria to be met, enabling certification and harmonization of characteristics and functionalities. For instance, ISO 23894, expected in 2025, establishes ethical principles for AI systems, such as algorithm transparency, data privacy, and fairness in outcomes. Another relevant standard is ISO 38505, which focuses on data governance; when applied to AI systems, it ensures that data is used ethically, fairly, and transparently. This is crucial in educational and healthcare contexts where the handling of sensitive data is a key component of processes and where individuals' rights must be respected (Arrieta, 2024).

At least, the findings underscore the importance of fostering interdisciplinary dialogue to address the ethical challenges posed by GenAI. The complexities of algorithmic bias, data privacy, and transparency require collaborative input from educators, technologists, ethicists, and policymakers. Without such collaboration, there is a risk that the ethical implementation of GenAI will remain an aspirational goal rather than a tangible reality.

Broadening this discussion, it serves to point out how the integration of Generative AI (GenAI) into Latin American higher education institutions exposes a wide spectrum of ethical and educational dilemmas that must be addressed to ensure equitable, effective, and inclusive implementation. This discussion delves deeper into the dilemmas, challenges, and unresolved questions, aligning them with institutional and societal imperatives.

Ethical and Educational Dilemmas

From the findings described in the literature review, the institutional documents, and the research on the teaching community in Mexico, Brazil, and Colombia, there are some points to note regarding the most relevant ethical and educational dilemmas (See Scheme 1).

The following tables synthesized some of the most important elements that we have mentioned, in terms of dilemmas and strategies (Tables 1 and 2).

Challenges and Questions

These dilemmas raise some questions and challenges (See Fig. 4). Asking the following questions will help to think prospectively about the elements that should be observed to enrich any dialogue, policy, or pedagogical action that may be proposed in diverse contexts of higher education.

The findings reveal that institutional readiness varies significantly across the region. For instance, while Brazil exhibits pioneering efforts in AI governance, Colombia and Mexico face challenges in aligning policy frameworks with ethical standards. This disparity necessitates harmonized and collaborative approaches to ensure that no institution or nation is left behind in the ethical integration of AI. Moreover, the tension between regulation and innovation must be resolved. Universities need to navigate these competing demands by fostering environments where ethical use of GenAI is incentivized and supported. Encouraging experimentation with AI while maintaining ethical oversight could provide a pathway forward.

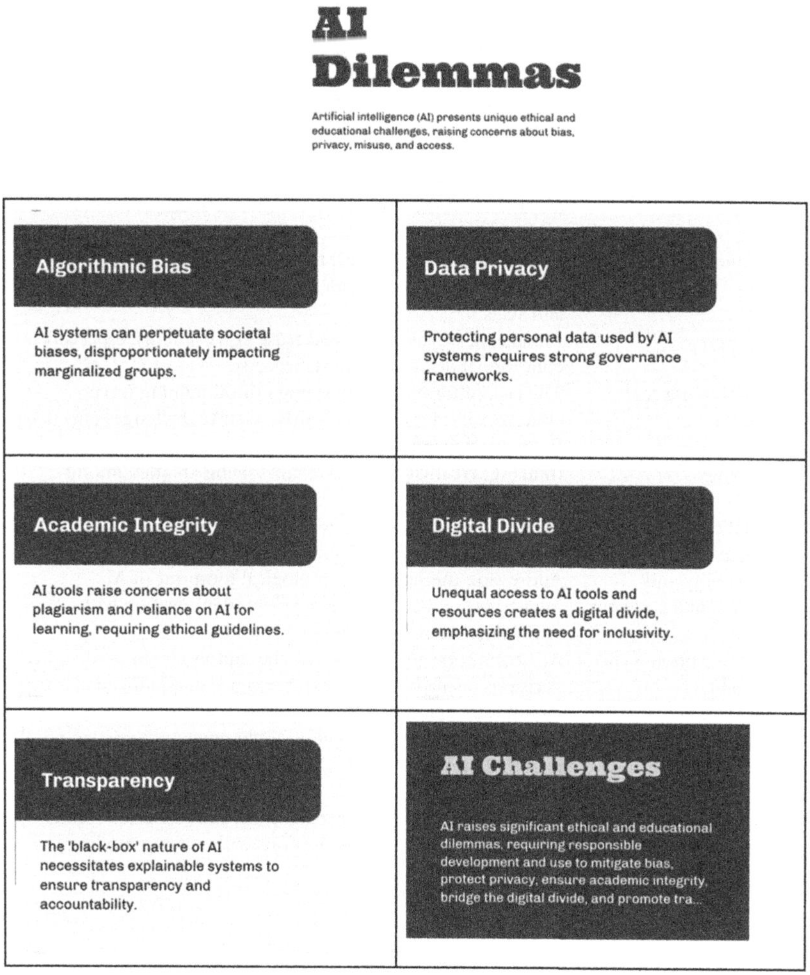

Scheme 1 AI dilemmas. Source. Authors

Conclusion

This chapter highlights both the opportunities and challenges inherent in the ethical implementation of generative AI within higher education institutions in Latin America. The potential of GenAI to revolutionize pedagogical practices, enhance research capabilities, and democratize access to

Table 1 Key ethical and educational dilemmas

Algorithmic bias	Addressing how AI perpetuates societal inequalities, especially affecting underrepresented groups in education
Digital equity	Bridging gaps in access to ensure inclusivity and reducing disparities in AI adoption across diverse socio-economic backgrounds
Privacy	Developing robust data protection mechanisms, and reducing disparities in AI adoption across diverse socio-economic backgrounds
Academic integrity	Balancing innovation with the prevention of academic misconduct, ensuring students develop authentic knowledge and skills
Transparency	Enhancing explainability and trust in AI systems, especially in educational decision-making processes
Ethical training for educators	The lack of formalized, ongoing ethical training leaves educators ill-prepared to navigate GenAI challenges effectively
Over-reliance on Technology	Excessive dependence on AI tools may undermine critical thinking, creativity, and problem-solving abilities among students
Cultural context and relevance	Ensuring AI tools are designed and implemented to respect and integrate cultural, linguistic, and regional diversity
Sustainability and environmental concerns	Addressing the significant ecological footprint of AI infrastructure, balancing technological innovation with environmental responsibility
Teacher autonomy and adaptability	AI-driven tools may restrict teacher autonomy in curriculum decisions or inadvertently enforce rigid, standardized educational practices
Misaligned objectives	The adoption of AI without clear alignment with pedagogical goals risks prioritizing efficiency over meaningful learning outcomes
Bias in educational resource allocation	Unequal access to AI-enhanced tools and resources could create divisions among institutions with varying funding capacities

Source. Authors

quality education is undeniable. However, these benefits are contingent upon the development of robust, context-sensitive ethical frameworks that address the unique socio-cultural and economic realities of the region.

The formulation of such frameworks must prioritize inclusivity and equity, ensuring that GenAI serves as a tool for empowerment rather than exclusion. This requires concerted efforts to bridge the digital divide, particularly for marginalized communities. Moreover, academic integrity and transparency must remain central pillars of any ethical guidelines, fostering trust among educators, learners, and society at large.

Table 2 Strategic recommendations

Regional harmonization of policies	A unified framework informed by international standards (like ISO 23894 and ISO 38505) but tailored to local contexts is essential. Collaborations like the Santiago Declaration and UNESCO's initiatives provide a foundation for such efforts (while emphasizing local socio-economic contexts and educational needs)
Equity-centric approaches	Efforts to democratize access to GenAI must prioritize marginalized communities, ensuring that technological advancements do not deepen existing inequities
Digital literacy initiatives	Universities must integrate digital literacy programs into curricula to equip students and faculty with the critical skills necessary to navigate AI technologies responsibly
Interdisciplinary collaboration	Ethical challenges require collaborative solutions that involve educators, technologists, policymakers, and ethicists working together to design innovative strategies. Create forums and working groups involving educators, technologists, ethicists, and policymakers to develop innovative, context-aware solutions to GenAI challenges in education
Sustainability and environmental awareness	The environmental impacts of AI systems, though currently underexplored, should form an integral part of future discussions and strategies
Teacher training and support	Establish continuous professional development programs for educators to enhance their understanding and application of AI in pedagogical settings, with a focus on ethical implications and adaptive methodologies
Strengthening academic integrity	Implement robust anti-plagiarism tools and promote a culture of ethics through workshops, institutional policies, and the integration of integrity-focused modules in educational programs
Customizing AI for regional needs	Invest in AI systems that reflect regional linguistic, cultural, and educational priorities, ensuring their relevance and acceptance within local contexts
Community involvement	Engage students, parents, and community stakeholders in discussions about AI adoption in education to ensure inclusive and transparent decision-making processes
Promoting ethical AI design	Encourage partnerships between universities and technology developers to co-create AI tools that align with pedagogical values, prioritize fairness, and minimize algorithmic bias
Enhanced resource allocation	Develop funding programs for AI implementation in resource-poor educational institutions, ensuring equitable access to advanced educational technologies across Latin America
AI literacy for leadership	Train educational administrators and policymakers to understand and guide AI integration within their institutions effectively, ensuring alignment with institutional missions and goals

Source. Authors

Fig. 4 Challenges and questions. Source. Authors

The study also underscores the need for a regional approach to AI governance. Harmonized policies and regulations, informed by international best practices yet tailored to local contexts, are essential for mitigating risks and maximizing the positive impact of GenAI. Latin America has the opportunity to position itself as a global leader in ethical AI implementation by leveraging its rich cultural diversity and addressing its systemic challenges with innovative solutions.

Ultimately, the ethical integration of GenAI in education is not merely a technological or institutional challenge—it is a societal imperative. By embracing interdisciplinary collaboration, fostering digital literacy, and committing to principles of justice and equity, higher education institutions in Latin America can shape a future where technology amplifies human potential and upholds the values that define our shared humanity. This vision, while ambitious, is both necessary and achievable with the collective efforts of all stakeholders involved.

Approaching a pedagogical goal in conducting this study it is important for the reader to keep in mind that the ethical implementation of GenAI in higher education is an urgent and multifaceted challenge that demands a cohesive response from institutions, governments, and society. This study illuminates the complexities of aligning AI technology with the principles of equity, transparency, and integrity in a region as diverse as Latin America.

The research highlights the need for continuous dialogue and adaptive strategies to align AI's transformative potential with educational values. By leveraging Latin America's rich cultural diversity and fostering inclusive innovation, the region has the opportunity to emerge as a global leader in ethical AI integration.

The ethical deployment of GenAI in education is a societal imperative that transcends technology. It demands a commitment to justice, equity, and human-centered values. Through collective efforts, Latin American institutions can redefine the role of technology in education, ensuring it serves as a catalyst for empowerment and transformative learning.

Acknowledgments The research presented in this chapter was conducted in collaboration with the RIDUL Group at La Salle Universities (Mexico, Brazil, Colombia), under the authorship of Rubén Yáñez, Paola Mercado, and Jairo Alberto Galindo. Key contributors, including Cecilia Vallejos, Anabell Covarrubias, Celina Gastellum, and Judite Sanson, provided invaluable insights that were instrumental in shaping the study. Their expertise and dedication ensured the research's alignment with the specific needs and challenges of the Latin American context, significantly enhancing its relevance and academic rigor.

References

Albuquerque, A. M., Reis, M. L., Bizelli, J. L., Soligo, M. D. C. A. L., & Miranda, N. A. (2023). Experiencias y propuestas sobre la organización y gestión universitaria de la IA en Brasil. In J. Gairín Sallán & L. Alguacil Mir (Eds.), *La gestión de la Inteligencia Artificial en los contextos universitarios iberoamericanos* (pp. 97–110). Edo-Serveis-Universitat Autónoma de Barcelona.

Andión Gamboa, M., & Cárdenas Presa, D. I. (2023). Convivir con inteligencias artificiales en la educación superior: Retos y estrategias. *Perfiles Educativos, 45*(Especial), 56–69. https://doi.org/10.22201/iisue.24486167e.2023.Especial.61691

Arce González, C., Rubio, J. G., & Carrillo, L. S. G. (2023). Experiencias y propuestas sobre la organización y gestión universitaria de la IA en Colombia. In J. Gairín Sallán & L. Alguacil Mir (Eds.), *La gestión de la Inteligencia Artificial en los contextos universitarios iberoamericanos* (pp. 128–139). Edo-Serveis-Universitat Autónoma de Barcelona.

Arrieta, F. (2024, diciembre 29). Normas ISO y la ética en inteligencia artificial. *Noticias.* Retrieved from https://noticias.perfil.com/noticias/empresas/normas-iso-y-la-etica-en-inteligencia-artificial.phtml

Benemérita Universidad Autónoma de Puebla (BUAP). (2024). *En la BUAP se aborda el impacto de la IA en la educación e industria editorial.* Retrieved from https://www.boletin.buap.mx/node/3616

Binns, R. (2018). Fairness in Machine Learning: Lessons from Political Philosophy. In *Proceedings of the 1st Conference on Fairness, Accountability and Transparency, in Proceedings of Machine Learning Research* (Vol. 81, pp. 149–159). Retrieved from https://proceedings.mlr.press/v81/binns18a.html

Blackwell, A. (2024). What are Large Language Models Good For? In M. V. Faul (Ed.), *AI and Digital Inequities. Policy Insights #04.* NORRAG.

Botero, C., & Sáenz, P. (2022). *Superar la brecha digital, la justicia social también es una cuestión de conectividad | Heinrich Böll Stiftung*. Retrieved from https://co.boell.org/es/2022/03/15/superar-la-brecha-digital-la-justicia-social-tambien-es-una-cuestion-de-conectividad

Cobo, C. (2016). *La Innovación Pendiente. Reflexiones (y Provocaciones) sobre educación, tecnología y conocimiento*. Colección Fundación Ceibal/Debate: Montevideo.

Copeland, B. J. (2024). History of artificial intelligence (AI). *Encyclopedia Britannica*. Retrieved from https://www.britannica.com/science/history-of-artificial-intelligence

Currie, G. M. (2023). Academic Integrity and Artificial Intelligence: Is ChatGPT Hype, Hero, or Heresy? *Seminars in Nuclear Medicine, 53*(5), 719–730. https://doi.org/10.1053/j.semnuclmed.2023.04.008

Dabis, A., & Csáki, C. (2024). AI and Ethics: Investigating the First Policy Responses of Higher Education Institutions to the Challenge of Generative AI. *Humanities and Social Sciences Communications, 11*(1). https://doi.org/10.1057/s41599-024-03526-z

European Union. (2024). *Configuring the Digital Future of Europe*. Retrieved from https://digital-strategy.ec.europa.eu/policies/regulatory-framework-ai

Floridi, L. (2019). *The Logic of Information: A Theory of Philosophy as Conceptual Design*. Oxford University Press. https://doi.org/10.1093/oso/9780198833635.001.0001

Gallent Torres, C., Zapata González, A., & Ortego Hernando, J. L. (2023). El impacto de la inteligencia artificial generativa en educación superior: Una mirada desde la ética y la integridad académica. *RELIEVE—Revista Electrónica De Investigación Y Evaluación Educativa, 29*(2). https://doi.org/10.30827/relieve.v29i2.29134

Instituto Politécnico Nacional (IPN). (2021). *Con inteligencia artificial politécnicos crean herramienta de enseñanza del idioma inglés*. Retrieved from https://www.ipn.mx/imageninstitucional/comunicados/ver-comunicado.html?y=2021&n=26

Jiang, Z., & Zhou, J. (2024). Ethical Considerations and Challenges of AI in Higher Education: Analysis from the Perspective of International Organizations. In *Encyclopedia of Educational Innovation* (pp. 1–6). https://doi.org/10.1007/978-981-13-2262-4_293-1

Jobin, A., Ienca, M., & Vayena, E. (2019). The Global Landscape of AI Ethics Guidelines. *Nature Machine Intelligence, 1*(9), 389–399. https://doi.org/10.1038/s42256-019-0088-2

Kuhl, P., et al. (2019). *Developing Minds in the Digital Age: Towards a Science of Learning for 21st Century Education, Educational Research and Innovation*. OECD Publishing. https://doi.org/10.1787/562a8659-en

LAIA. (2024). *Latin American Artificial Intelligence Index 2024*. Retrieved from https://indicelatam.cl/wp-content/uploads/2024/12/ILIA_2024_compressed.pdf

Levinson, M., & Fay, J. (2016). *Dilemmas of Educational Ethics: Cases and Commentaries.* Harvard Education Press. Retrieved from https://eric.ed.gov/?id=ED568736

McMahon, E., & Milligan, L. O. (2021). A Framework for Ethical Research in International and Comparative Education. *Compare: A Journal of Comparative and International Education, 53*(1), 72–88. https://doi.org/10.1080/03057925.2021.1876553

Medrano Rodríguez, H., Morales Artero, J. J., de Sollano, D., Núñez, N., Guerrero Rodríguez, L., & Medina Morales, G. (2024). Experiencias y propuestas sobre la organización y gestión universitaria de la IA en México. In J. Gairín Sallán & L. Alguacil Mir (Eds.), *La gestión de la Inteligencia Artificial en los contextos universitarios iberoamericanos* (pp. 176–191). Edo-Serveis-Universitat Autónoma de Barcelona.

O'Neil, C. (2016). *Weapons of Math Destruction: How Big Data Increases Inequality and Threatens Democracy.* Crown Publishing Group.

Ożegalska-Łukasik, N., & Łukasik, S. (2023). Culturally Responsive Artificial Intelligence—Problems, Challenges and Solutions. arXiv preprint arXiv: 2312.08467.

Ramírez, A., & Casillas, M. A. (2024). Percepciones docentes sobre la Inteligencia Artificial Generativa: El caso mexicano. *Revista Paraguaya de Educación a Distancia, 5*(2), 44–55.

Reuters. (2024, July). *Brazil Announces $4 Billion Investment in AI to Achieve Technological Autonomy.* Retrieved from https://www.reuters.com/technology/artificial-intelligence/brazil-proposes-4-billion-ai-investment-plan-2024-07-30/

Rodríguez, M., Silva, R., Borges, A. P., Franco, M., & Oliveira, C. (2024). Artificial Intelligence: Threat or Asset to Academic Integrity? A Bibliometric Analysis. *Kybernetes, 53*(4), 1234–1256. https://doi.org/10.1108/K-09-2023-1666

Schmidt, S. (2024). Universidades brasileiras discutem regras de uso de inteligência artificial. *Revista Pesquisa.* Retrieved January 15, 2025, from https://revistapesquisa.fapesp.br/universidades-brasileiras-discutem-regras-de-uso-de-inteligencia-artificial/

SIPSE. (2024). *La brecha digital: El desafío pendiente para un México conectado.* sipse.com. Retrieved from https://sipse.com/tecnologia/la-brecha-digital-desafio-pendiente-para-un-mexico-conectado-483114.html

Sotheby's. (2024). A.I. God. Portrait of Alan Turing. *En Digital Art Day Auction.* Retrieved from https://www.sothebys.com/en/buy/auction/2024/digital-art-day-auction-2/a-i-god-portrait-of-alan-turing

Tauginienė, L., Gaižauskaitė, I., Glendinning, I., Kravjar, J., Ojstersek, M., Robeiro, L., Odineca, T., Marino, F., Cosentino, M., Sivasubramaniam, S., & Foltynek, T. (2018). *Glossary for Academic Integrity.* ENAI Report (revised version). Retrieved from https://www.academicintegrity.eu/wp/wp-content/uploads/2023/02/EN-Glossary_revised_final_24.02.23.pdf

Tesar, M., Peters, M., & Jackson, L. (2021). The Ethical Academy? The University as an Ethical System. *Educational Philosophy and Theory, 53*(5), 419–425. https://doi.org/10.1080/00131857.2021.1884977

Tubella, A. A., Mora-Cantallops, M., & Nieves, J. C. (2023). How to Teach Responsible AI in Higher Education: Challenges and Opportunities. *Ethics and Information Technology, 26*(1). https://doi.org/10.1007/s10676-023-09733-7

Turing, A. M. (1950). Computing Machinery and Intelligence. *Mind, 59*(236), 433–460. https://doi.org/10.1093/mind/LIX.236.433

UNESCO. (2019). *Beijing Consensus on AI and Education.* Retrieved from https://www.unesco.org/es/articles/la-unesco-ha-publicado-el-primer-consenso-sobre-la-inteligencia-artificial-y-la-educacion

UNESCO. (2021). *Recommendation on the Ethics of Artificial Intelligence.* Retrieved from https://unesdoc.unesco.org/ark:/48223/pf0000380455_spa

UNESCO. (2023). *Guidance for Generative AI in Education and Research.* Retrieved from https://unesdoc.unesco.org/ark:/48223/pf0000386693

Universidad Autónoma de San Luis Potosí (UASLP). (2023). *Crea UASLP grupo institucional para análisis de retos y oportunidades del uso de la Inteligencia Artificial en la educación.* Retrieved from https://wp.uaslp.mx/noticias/investigacion/crea-uaslp-grupo-institucional-para-analisis-de-retos-y-oportunidades-del-uso-de-la-inteligencia-artificial-en-la-educacion/

Universidad Autónoma Metropolitana (UAM). (2024). *Celebrará la UAM el foro académico "Perspectiva multidisciplinaria de la IA desde la educación superior".* Retrieved from https://50aniversario.uam.mx/celebrara-la-uam-el-foro-academico-perspectiva-multidisciplinaria-de-la-ia-desde-la-educacion-superior/

Universidad de Guadalajara (UDG). (2023). *Guía para el uso de la inteligencia artificial generativa en procesos académicos.* Retrieved from https://www.udg-virtual.udg.mx/Guia-IA-UDG

Universidad Nacional Autónoma de México (UNAM). (2023). *Recomendaciones para el uso de la inteligencia artificial generativa en docencia.* Retrieved from https://cuaed.unam.mx/descargas/recomendaciones-uso-iagen-docencia-unam-2023.pdf

The Quest for a Responsible Use of AI in Latin America

Rosa María Alonzo González
and David Ramírez Plascencia

Introduction

This volume focused on the actual discussion about how governments, corporations, and users could responsibly and safely integrate artificial intelligence into their daily activities and strategic actions without compromising human rights and environment. However, considering the ethical and legal issues related with disruptive technologies, such as the case of AI or robotics, it is not an easy task. This is mainly because there is a huge

R. M. Alonzo González (✉)
Facultad de Ciencias Humanas, Universidad Autónoma de Baja California, Mexicali, Baja California, Mexico
e-mail: rosa.alonzo@uabc.edu.mx

D. Ramírez Plascencia
Centro Universitario de Guadalajara (CUGDL), Universidad de Guadalajara (UdeG), Guadalajara, Jalisco, Mexico
e-mail: davidram@udgvirtual.udg.mx

D. Ramírez Plascencia, R. M. Alonzo González (eds.), *Ethical and Legal Dilemmas of Artificial Intelligence in Latin America*,
https://doi.org/10.1007/978-3-031-86540-4_9

gap among the existing regulations and the popularization of AI devices and applications among the public. Apps such as AI chatbots are becoming popular not only among students or corporations but in the public administration as well. This, though, without the formulation of minimum legal guidelines about privacy or liability. Although it is possible to mention the existence of general AI ethical frameworks created by international organizations and adopted by Latin American countries in their AI national strategies and some normative projects. Those principles need to land in concrete normative codes that provide clear and operative directives that norm the services and use of these systems and the devices that operate with them. This fact becomes even more complex considering regions, such as Latin America, whose legal frameworks usually are outdated and their authorities lack the infrastructure and resources to establish efficient protections to users, who are exposed to scams, the filtration of their personal data and the unauthorized uses of their personal information to create AI content.

Traditionally, there has been a breach between technology and regulation. It happened with P2P music interchange in early 2000 and the case of Napster that changed traditional intellectual property regulations and, actually with the cryptocurrency markets that many times operate under the radar of countries' regulations. The process of adoption of AI in Latin America, as it was possible to observe in this volume, followed the same pattern as other technological advances such as social media or drones. While Latin Americans are already avid users of AI chatbots such as ChatGPT, just in early 2023, there were counted 303 million visits from this region (Salazar Castellanos, 2023) and the economic projection foresees an exponential grown in the regional chatbot market from 596.2 million in 2023 to reach a revenue of US $2398 million by 2030 (Grand View Research, 2024) However, As observed in Chap. 6, this expansion contrasts with the slow, and sometimes inefficient, legislative processes. Besides in the last years, there have been several reports and AI national policies, those documents have served only as *good intention letters* that make emphasis on promoting a responsible operation of AI. Only a few countries as the case of Chile or Brazil have advanced significantly in developing proper legal frameworks that support not only the development of AI but also to guarantee that these technological advances follow ethical guidelines based on the protection of users' rights and the supremacy of public interest over corporative revenues.

In recent years, as the presence of AI grows in Latin America and its employment expands into the economic, media and governmental sectors, some concerns, controversies and even punctual cases have emerged regarding the establishment of ethical limits, the liability of developers and users and the role of the authorities to regulate AI. This volume focused on the analysis of some of these recent controversies and problems caused by the accelerated adoption of AI applications in the media, security, migration, government administration, and educational and economic activities in the region. This volume also concentrated on highlighting the potential benefits but also the patent risks of incorporating AI without having efficient legal and ethical frameworks.

One of these key controversies is the impact of AI in the media sector and journalism, as digital content creators, newspapers and journalists are already employing, or considering- using AI apps to develop multimedia and news. AI is bringing great advantages such as providing faster content and reducing the workload of performing repetitive tasks. Journalists and social media creators could spend those efforts on other substantial actions such as doing research or preparing the information. But the introduction of AI may also affect the already adverse labor situation of media workers (Neuman, 2024), as some of their activities could eventually be replaced as the AI systems continue improving. These worries about a potential forthcoming shift in the labor market because of the constant growth of AI in Latin America are not only present in the media and entertainment sectors, as AI expands into other economic sectors (Tata Consultancy Services, 2024). The incorporation of AI may also affect freelance workers and Internet sellers that work across online platforms, and whose reputation, rating and popularity schemes depend not necessarily on quality, but sometimes on directives about cost-benefits settled by corporations through the algorithms that operate those platforms. Regarding the labor market and AI, it is pertinent to ask some key questions: how will this technology affect workers' rights? In what way will the government and the corporations face legal issues regarding intellectual property rights of content generated by AI? Or the liability of potential misuses, who will be responsible?

Another important theme is related with the transformation of security and homeland protection across the novel AI devices and systems, some already present, such as the case of drones and IA cameras and face recognition technology employed to secure borders, and others in development as the case of crime-prevention systems. There are pending questions about the legal and ethical principles that sustain the operation of drones or surveillance cameras particularly regarding the presumption of innocence and the right of privacy of citizens, especially regarding people in a situation of high vulnerability as the case of migrants. This is imperative considering that many of the AI systems are operated by foreign corporations who could potentially gather strategic data from devices used by humanitarian organizations. It is also mandatory to set protocols to avoid collateral damages or victims caused by armed drones (UCAVs), as these devices are becoming more common in Latin America.

The impact of AI education is another relevant issue in the public debate in the last years. It is not only about the promising tools that would eventually enhance pedagogic strategies but also about generative AI applications that allow students to explore and create their own content. But this innovation has bright concerns mainly regarding the misuse of these apps when doing homework. Then the education systems are paying attention not only to the benefits but also to implementing those systems in a responsible way. Here, it is important to consider essential questions such as how to construct ethical guidelines to promote the responsible use of AI in the classroom? How could teachers promote autonomy but also to entrust students about the correct use of these apps?

Volume Key Findings

This volume approached the analysis of the AI adoption in Latin America from two complementary perspectives. Volumes of Part I, "Considering AI from the Private Sphere," pondered this issue from the perspective of citizens and corporations. In what way individuals and organizations employ artificial intelligence systems in the region? Part II, "Challenges and Promises of AI in the Public Sector," on the contrary, focused mainly on policies already established or that are on the transition of being applied to regulate the use of AI in issues related with the public interest. Though this division among the private sphere and the public sector is just for didactic purposes as some actions, such as the case of introducing AI security cameras in public spaces or licensing AI education software at public schools, convey responsibilities and rights for the government, the users, and the developers.

Part I delivers compressive legal and ethical analysis on experiences and problems related with the introduction of AI to potentialize the media and the labor market, but also focused on understanding what kind of barriers and controversies have appeared among the corporations and the employers, and in what way AI has the capability to bring innovation but also the introduction of these systems in the media sector, for example, may cause labor precarization and the loss of employments, as some activities are being replaced by automatization. Chapter 2, "Artificial Intelligence, Intellectual Property, and the Labor Precarization of Journalists in Latin America," by David Ramírez analyzed, through a compressible study, the introduction of AI in the media sector, specifically journalism. How journalists in Latin America approach AI to improve their work: from achieving research to increasing the quality and speed in the news production process. But also, in what way the adoption of this technology may worsen the already adverse labor situation among journalists, as AI systems are becoming more popular, not only to manage information but to forge content from images, graphics, and news. However, as the author affirmed, the economic and legal impact of AI in the journalism and media sector cannot be seen as a disrupted and unprompted event, but as part of a long process of digital media adoption that has relentlessly changed the media business over the past three decades. As journalists and newspapers in the region have been compelled to reduce fixed costs and incorporate computers, social media, and now AI, to comply with a public more avid to consume content instantly and lower or rid costs.

Considering technology adoption as a societal process, their impact is many times shaped by how public perceptions embedded this technology with their own preconceptions and concerts. The case of AI is paradigmatic, because many times, the public formulated a galore of imaginaries and idealization based on popular culture, comics, or movies, or throughout how the media portrays this technology. In Chap. 3, "From Epistemological Foundations over Futuristic Speculations to Fact-Based Concerns: Evolving Discussions on Artificial Intelligence in Argentina, Colombia, and Mexico (2014–2024)," Fátima Ávila-Acosta, Lucía Morales Lizárraga, Jan Nehring approached the problem of how Latin America media frames artificial intelligence. In this work, they made an extensive analysis on how newspapers provide encountered visions where there is a convergence of a positive foresight on the potential benefits but also warnings and cautionary reflections on the lack of regulation and their negative impacts. These conclusions are very relevant considering that in many ways recent normative proposals and policies in Latin America mirror these opposed visions on AI. On one side, enthusiasm about how AI may improve countries' economies and people's lives, but on the other side, many bills consider the introduction of AI systems risks, surveillance office and liability schemes. There are also frequent remarks on the predominance of human rights over AI performance, which from an ethical perspective is an adequate approach but it also shows apprehensions about future and potential misuses, even regarding systems that are just only on consideration.

As a matter of fact, one of the key topics regarding the employment of AI in the productive sectors in Latin America, in Chap. 2, it was possible to observe how automation and generative AI could cause, in the long-run, labor precarization and the loss of jobs regarding the news sector, but this tendency could expand into other creative industries, as the AI systems become more efficient and the industry employs them not only to manage information flows, but also to create content. Chapter 4, "The Algorithmic Tyranny in the Gig Economy: National Strategies and Policy Implications in the Latin American Context," approached this issue but it focused on how national strategies impact employment, particularly the digital labor in the region. Alisa Petroff started her chapter with an analysis of the technological background. She identified the strategic information technologies areas such as the case of high-speed Internet, that must be reinforced and updated to enable the development of AI systems and to avoid persistence structural inequalities. Then, the author made a comparative analysis of the national strategies that show their opportunities and weaknesses. In the following sections, Petroff concentrated on the most important impacts of AI in the labor market, the potential economic prospective profits and how the AI would introduce new jobs in the gig economy. However, in the final conclusions, the authors assert that as the presence of AI in the region's economies grows, there will be some challenges to gig workers. One of them is to adapt to the changes in the systems of evaluation that validate workers' reputation. As these systems increasingly rely on algorithmic logic without human intervention, the challenges will increase as the global economy stands more on outsourcing and the hire of freelancers through labor platforms that operate more under a dynamic of high productivity and low cost. This phenomenon is already present, for example, in the outsourcing of annotating data for artificial intelligence (AI), which usually operates in companies established in poor countries whose workers labor in precarious conditions (Muldoon et al., 2025).

In the final work in this section, Chap. 5, "Mapping the Future of AI Regulation in Latin America: Civil Society Perspectives on Brazil's Pilot AI Regulatory Sandbox," Kenzo Soares Seto studied the civic engagement in the establishment of AI regulation in Brazil. Soares focused on the case of legal proposals submitted by public organizations to attend the invitation of the Brazilian National Data Protection Agency (ANPD) to develop a regulatory sandbox for Artificial Intelligence (AI). In a compressive work, the author identified the diverse strategies and agendas of the activist organizations that participated in the public consultation in 2023. The objective of the ANPD was to promote the public collaboration of diverse agents such as individuals, social organizations, scholars and the private sector to create and test, through the figure of a sandbox regulation, future AI regulations that endorsed algorithmic transparency and responsible innovation. Soares Seto's study delivered an insightful analysis from Brazilian experience about whether it is possible to open the discussion of AI regulation to the civic organizations, enriching the discussion and promoting public opinion's validation. These civic participation exercises enable people to express not only their worries but also their expectations about the present and future of AI and is also helpful to foresee potential legal flaws before approving the new normativity, avoiding prospective legal controversies and gaps.

Volume's Part II. Challenges and promises of AI in the public sector, delivered comprehensive studies about the ethical and legal features caused by the introduction of artificial intelligence systems in key public areas: public administration, education, security, and border control. In Chap. 6, "Ethical and Legal Dilemmas of National AI Policies in Latin American Countries," Alonzo González made a substantial analysis of the national policies and legal bills developed in Latin America in the last years. In Alonzo's chapter, it is possible to observe the historical process of establishing a national framework that could guide the adoption and development of AI, but that also offers legal certainty to developers, investors, and users about the rights, liabilities, and potential risks derivative from the operation of devices. But as the author asserted, the process of founding these general guidelines has been lagging and insufficient. As it has happened traditionally with the regulation of other technologies such as bitcoins or social media, there is a patent timing gap between artificial intelligence popularization and employment in both the public and private sectors and the official paths to set ethical and legal frameworks. People, governmental officials, teachers, students, and corporations are already using AI applications to create multimedia content, chatbots to get information and in some cases AI cameras and sophisticated drones. But these practices have been occurring without proper legal frameworks and technical procedures that guarantee essential rights such as privacy and the protection of sensitive data, particularly in the case of public agencies that manage medical, homeland security and financing records. Another problem depicted in Chap. 6 is the lack of secondary regulations that offer operational guidelines. While most of the countries in Latin America have been working on general policies and regulations, it is evident the absence of secondary laws that provide technical rules and procedures that guide developers, officers and users. For example, most of the national strategies established that AI systems must protect users' privacy, but only few provide methods on how this right would be guaranteed and which agency would supervise the correct application of AI systems.

Next chapter, Chap. 7, "The Use of Biometrics and AI for Border Control and Its Impact on Human Rights," by Jezabel Pérez and Natalia Brzezinski stood on the interception amid security, AI and the protection of human rights. In recent years, authorities have incorporated AI systems and smart devices to increase homeland security from terrorist attacks, as the case of the US, or to combat the rising internationalization and expansion of criminal organizations across the region (the Mexican drug cártels and the Venezuelan gang called "El Tren de Aragua"). These Criminal gangs have diversified and improved their operations thanks to the introduction of sophisticated technology such as drones to surveillance labor and more recently to increase their firepower and cryptocurrencies for money-laundering, the AI cameras and other AI systems would eventually also include into their resources. The authors focused on how the United States and other Latin American countries incorporate emerging technology such as AI to analyze biometric and personal data from migrants. Besides, the justification of employing these technologies to increase national security to combat narcotraffic and transnational crime, stands as a flagrant violation of migrants' human rights who are compelled to provide private data without warrants or certitude about the future employment of this information. As authors asserted in their conclusion, there should be a balance among the protection of the country against external menaces, the incorporation of new technological advances, as the case of drones, AI, biometric and facial recognition, and the protection of human rights. This last fact is imperative considering the international and national commitments to promote technological development following ethical and legal frameworks centered on human protection.

In Chap. 8, "Ethical and Educational Dilemmas of AI in Latin American Higher Education Institutions: Persistent Challenges and Inquiries," Jairo Alberto Galindo-Cuesta, Rubén Yáñez, and Paola Mercado joined the key discussion about using AI apps to innovate in the teaching-learning process. This has been the case of applications such as ChatGPT, a chatbot that users can interact in human-like dialogue and is able to accomplish diverse tasks such as create essays, opinions, and definitions. As users, particularly teachers and students, become more familiar using Generative AI to create education content and do schoolwork, there are growing concerns about potential misuses of these tools, such as cheating at school and presenting exercises as being created by their own but completely made by AI. In this case, no advantage or innovation derives from the application of AI, but it may cause a detriment in students' learning. The chapter examined how Latin American universities are dwelling with potential benefits but also the ethical issues caused by the popularization of generative AI in classrooms. A discussion that is mandatory to enhance the research which faces its own dilemmas before the rising employment of AI in the academic production. As authors assert, the formulation of ethical guidelines must depart from the identification of key elements associated with the AI potentialities but also with the challenges that implies its use in the education sector. What kind of ethical frames will fit the most to encourage innovation in the traditional classroom but also in what way AI could be employed as a partner not a replacement of humans in academic research.

The Quest for a Responsible Use of AI

Across the years, there has been a recurrent topic in academic literature and public documentation related to AI in Latin America. This is the responsible use of AI (Pombo et al., 2020; Rodríguez, 2022; Cumbre Ministerial, 2024). Cambridge (2024) defines responsible as "to have control and authority over something or someone and the duty of taking care of it, him, or her." This definition implies two main topics: (a) to have control or authority and (b) to have a duty. This, in terms of regulation, relates with another issue, (c) liability. That is a "to be held legally responsible or obligated" (Wex Definitions Team, 2023). Therefore, to make a responsible use of AI links mainly with having the authority, in terms of an established legal framework, or to have control. Here the term control should be considered under a wider scope since the AI developers, the government, corporations and individuals have a certain degree of agency and control when developing or employed AI systems. Duty and liability could be considered from a legal and ethical perspective. Considering the legal aspect of duty and liability is a complex task, since many times both concepts are taken as equal, even legal codes sometimes employ these terms as interchangeable. However, they refer to slightly different things. Duty denotes someone that is compelled by law to undertake an action such as pay a debt or a tax (Wex Definitions Team, 2024), while the liability is an obligation established by law (IFRS, 2013). While both terms relate among them, duty denotes the obligated subject and liability is an abstract obligation established in the law, that only particularizes when there is a legal act such as a trial, the commission of an illegal act or the sign of contract etc. On the ethical aspect of duty and liability, should be considered depending on the philosophical perspective because there are diverse approaches on the subject (Zimmerman, 2013). Duty, in Kantian terms, is a categorical imperative, that is a moral rule that compels people to act according to universal laws. Bestowing to this principle, people must behave under a sense of obligation, no matter their desires or circumstances (Kant, 1997). Though, facing the discussion on the legal and ethical matters about duty and liability is far beyond the scope and material possibilities of this work, it is important to set that considering the ethics of AI, it embraces not just what is written on the codes but also those principles that support people's coexistence and relations, some of them do not necessarily comply with formal regulations. Here for example: the governments may set general guidelines that are not formally established in legal codes, and the developers could incorporate some inner limitations regarding their own systems, but also the individuals who interact with AI and use these apps to create content based on their own moral principles. However, considering the regulation of information technologies, this moral contra-position usually breaks out when human rights and the protection of users' spheres clash with media corporations' profit objectives and governments' surveillance policies (Piccolo, 2020).

Establishing a responsible environment to develop and use AI systems requires the cooperation of all actors: the government (at local, national and regional level) who should encourage a participative development of a legal framework that bring certitude to all actors, the developers should focus not only on the performance and revenues but also in protecting their users, and users (including individuals, corporations and the government) should incorporate these systems considering not only the potential advantages but also the duties, the responsibilities and the prospective liabilities inherent to the employment of this technology. In this context, it is mandatory that the legal and ethical frameworks of duties and liabilities should be established, but that those normative frameworks should not only mention the principles but also the technical and legal operational norms. While the development of legal normativity in Latin America is a pending question that must be fulfilled to encourage the expansion and cooperation of AI at a regional level, the ethical aspect of technology is equally important. All actors involved in the development of AI should be conscious of the importance and centrality of digital technology not only in the public and economic aspects, but also in the private sphere. Therefore, their actions may have a deep impact on the private life of individuals, affecting not only their economy but also their health. It has become recurrent, for example, the employment of generative AI apps to commit misuses by altering original images of minors. The implications of those conducts are beyond the legal aspects of formal regulation but imply the ethical duty of making a reasonable employment of technology. Promoting responsible use is not, however, a government exclusive duty, but a social undertaking that requires all actors' efforts. Endorsing a responsible development, regulation and use of AI will enable the forging of synergies, and the construction of an ambient of trust among all actors, along with the creation of a sustainable model that maximizes the potentialities of AI and diminishes the environmental externalities associated with the process of technological shifts such as e-waste and the actual impact on energy and water supply derived from the operating of AI data centers.

Final Remarks

In Latin America, the promotion of a responsible use of AI stands on the improvement of social and economic conditions, then the users could enjoy better technological infrastructure to take advantage not only on fresh new advances as the case of AI, but also on already established information technologies that require optimal computers systems and high-speed Internet. Users also need the development of digital competences that not only allows them to enjoy digital media for entertainment or communication but to transform their reality. Latin Americans need to become creators not only consumers, incorporating these technologies to widen their voices not just to comply with their private needs but also to express communal and social agendas that sometimes are obscured by international corporate and governmental interests from the Global North. It is imperative to diversify the sources of information from which AI systems are fed, to avoid the imposition of ideological dominance, but at the same time, it is mandatory to diminish digital extractivism that keeps traditional asymmetries.

Latin American governments should consider the artificial intelligence rising popularity as an opportunity to improve all the digital media legal ecosystem, that means to apply an extensive review of legal frameworks that regulate Information Technologies (IT): including the Internet, social media, mobile applications and smart devices among other technological gadgets, platforms and systems in which stand the actual development of AI. The participative construction of IT regulation is also important. This not only implies the development of national policies or the subscription to ethical frameworks but also the public participation on the composition of secondary regulations that sets the operational rules and technical specifications to develop, commercialize, use and apply AI systems in the private and public scope. There are key legal issues associated with the development and use of IT that must be attended to: (a) intellectual property law, privacy and data protection, censorship and free speech, and antitrust law and finally, but no less important, to improve accountability and the combat of corruption. Governments also need to encourage the awareness among users and developers of a core of ethical principles that promote the responsible use of information technologies, including AI. This should be especially focused on young people and minors, a sector who is particularly exposed to bullying and other digital misconduct. The promotion of a safe use of AI is equally important among corporations, the public administrations, and citizens, not only to protect sensible private data while using these devices but also not to affect other people's rights. Finally, in the consolidation of a responsible development of AI at a regional level, it is imperative to promote collaboration to set regional ethical and legal frameworks, incentivize private and public investments to build AI laboratories and data centers but also to raise the voice in international forums to incorporate regional needs in the multilateral agenda.

References

Cambridge. (2024, December 25). Responsible. *Cambridge Dictionary*. Cambridge Dictionary. Retrieved December 28, 2024, from https://dictionary.cambridge.org/dictionary/english/responsible

Cumbre Ministerial. (2024, August 9). Declaración de Cartagena de Indias para impulsar el uso de la IA con ética y responsabilidad.

Grand View Research. (2024). *Latin America Chatbot Market Size & Outlook, 2030*. Retrieved December 19, 2024, from https://www.grandviewresearch.com/horizon/outlook/chatbot-market/latin-america

IFRS. (2013). *Conceptual Framework: Definition of a Liability*. London. Retrieved December 28, 2024, from https://www.ifrs.org/content/dam/ifrs/meetings/2013/january/iasb/conceptual-framework/ap9b-definition-of-liability.pdf

Kant, I. (1997). In M. J. Gregor (Ed.), *Immanuel Kant Critique of Practical Reason* (1st ed.). Cambridge University Press. https://doi.org/10.1017/CBO9780511809576

Muldoon, J., Cant, C., Graham, M., & Ustek Spilda, F. (2025). The Poverty of Ethical AI: Impact Sourcing and AI Supply Chains. *AI & SOCIETY, 40*(2), 529–543. https://doi.org/10.1007/s00146-023-01824-9

Neuman, N. (2024). *Journalism, Media, and Technology Trends and Predictions 2024*. Reuters Institute. Retrieved December 18, 2024, from https://reutersinstitute.politics.ox.ac.uk/journalism-media-and-technology-trends-and-predictions-2024

Piccolo, F. L. (2020). Antigone, Today: Dignity and Human Rights in Contemporary Spaces. *Transactions of the Association of European Schools of Planning, 4*(1), 1–6. https://doi.org/10.24306/TrAESOP.2020.01.001

Pombo, C., Cabrol, M., Alarcón, N. G., & Ávalos, R. S. (2020). *fAIr LAC: Responsible and Widespread Adoption of Artificial Intelligence in Latin America and the Caribbean*. IDB. Retrieved July 24, 2023, from https://publications.iadb.org/en/fair-lac-responsible-and-widespread-adoption-of-artificial-intelligence-in-latin-america-and-the-caribbean

Rodríguez, M. (2022). *Uso responsable de la inteligencia artificial en el sector público* (Working Paper). Claudia Flores, Nathalie Gerbasi, María Isabel Mejiía, Martha Rodríguez y Antonio Silveira. Retrieved November 11, 2024, from https://scioteca.caf.com/handle/123456789/1941

Salazar Castellanos, D. (2023, April 11). ChatGPT: Which Latin American Countries Are the Biggest Users of the AI Tool? *Bloomberg Línea*. Retrieved December 19, 2024, from https://www.bloomberglinea.com/english/chatgpt-which-latin-american-countries-are-the-biggest-users-of-the-ai-tool/

Tata Consultancy Services. (2024). *From Potential to Performance by Design: TCS AI for Business Study*. TATA.

Wex Definitions Team. (2023, June). Liable. *LII/Legal Information Institute*. Cornell Law School. Retrieved December 28, 2024, from https://www.law.cornell.edu/wex/liable

Wex Definitions Team. (2024). Legal duty. *LII/Legal Information Institute*. Cornell Law School. Retrieved December 28, 2024, from https://www.law.cornell.edu/wex/legal_duty

Zimmerman, M. J. (2013). Duty and Obligation. In *International Encyclopedia of Ethics*. Wiley. https://doi.org/10.1002/9781444367072.wbiee158

Index

A

Academic dishonesty, 146
  concerns about plagiarism and over-reliance on AI tools for assignments, 146
Academic integrity, 145, 151, 153, 158, 159
  challenges in maintaining ethical practices with GenAI; guidelines for addressing misconduct, 155; role in higher education policy, 147, 148; strategies for fostering creativity while preserving integrity, 151
Access to AI technologies, *see* Digital divide
Administration, 125, 127
Administrative, 127, 128
  See also Administration
Agence France-Presse (AFP), 24, 25, 30
AI development centers, 122
AI laboratories, 122
  See also AI development centers
Algoactivism, 68
Algorithm, 126, 130
  See also Software
Algorithmic Accountability Act, 69
Algorithmic bias, 145, 155, 158
  ethical implications in education; addressing biases in AI systems, 145, 158; impact on marginalized groups, 145
Algorithmic transparency, 92, 93, 96–97, 99
Alphabet (Google), 24, 27, 30, 108
  See also Google News
Apple, 108
Apple SIRI, 10
Argentina, 25, 31, 70–73
Artificial intelligence (AI), 125–137, 144–150, 152–157, 160
  chatbots, 166

D. Ramírez Plascencia, R. M. Alonzo González (eds.), *Ethical and Legal Dilemmas of Artificial Intelligence in Latin America*, https://doi.org/10.1007/978-3-031-86540-4

Artificial intelligence (AI) (*cont.*)
definitions and categories, 149
developers, 120, 122
ethical challenges; governance frameworks, 147, 155, 156, 160; transparency and explainability, 158
security cameras, 168
Artificial intelligence governance
harmonization with international standards, 155
national and regional frameworks, 155
See also Regulation of AI
Associação Brasileira das Empresas de Software (ABES), 90, 93–97, 99, 100
Asylum, 126, 127, 129, 130, 134, 136
Auditable mechanism, 87
Authority, 166, 167, 172, 173
Authorship, 23, 30, 31
Automatization, 22, 27, 32
Autoridade Nacional de Proteção de Dados (ANPD), 85–87, 89, 91–93, 95–100

B

Biometric, 125–137
Bitcoins, 171
Bolivia, 25
Border, 125–137
See also Borderline
Brazil, 8, 10, 13, 26, 31, 71, 73, 126, 129, 131, 132, 135, 136, 166

C

Case selection
Argentina, 44–46
Colombia, 44–46
Mexico, 44–46
Censorship, 25
Chatbots, 2, 4, 10
ChatGPT, 2, 11, 14, 108, 110, 166, 173
Chávez, Hugo, 25
Chile, 70, 71, 73–74, 88, 92, 166
China, 2, 4
Cobalt, 109
Collect/collection (as for personal data), 126, 129, 131, 133, 134
Colombia, 25, 26, 71, 74–75, 91, 92
Control, 125–137, 171–173
border control, 125–137
migration control, 127, 130, 136
Copyright infringements, 21, 30
See also Fair use
Cortana, 10
COVID-19 pandemic, 112
Crime-prevention systems, 167
Cryptocurrency, 166, 172
Cultural relevance of AI, 158
importance of adapting AI systems to linguistic and regional diversity, 158
Customs and Border Protection (CBP), 128, 130, 136
See also DHS

D

Data privacy and protection, 155
Data protection, 72–77
Democratic, 22, 23
Developers, 87, 88, 90, 94, 95
DHS, 126, 134, 135
Digital divide, 146, 154, 158
challenges in equitable access to AI technologies, 146, 152
recommendations for bridging gaps, 158
Digital extractivism, 175
Digital literacy, 88, 99, 159
integration into curricula; student skills development, 159

Discrimination, 86, 97, 130–132, 135
Drones, 4, 107, 111, 122
See also UCAV
Duty, 173–175

E
Ecuador, 71, 75
El País, 25, 30
Equity in education, 145
ensuring inclusive practices with AI tools, 146
See also Digital divide
Ethical, 129–133, 135
Ethical AI guidelines, 144, 150, 158
implementation in educational institutions, 149
role in regional declarations and international standards, 159
EU Platform Work Directive, 69
European Unión, 2, 6, 7

F
Facebook, 108
Face recognition (technologies), 167
Fair use, 29
Fixed costs, 24, 26, 29
Freelancers, 22, 26, 32

G
Generative artificial intelligence (GenAI), 5, 10, 13, 22, 27, 29, 144–153
overview and applications; definitions and capabilities, 146, 148; ethical implications in teaching and learning, 147
Gig economy, 170
See also Gig workers
Gig workers, 67, 170
Global North, 7, 8
Global South, 5, 85–88, 94, 100, 101
Google, 108
Google News, 30
Governments, 125, 127, 128, 130, 131, 134, 136
Guatemala, 25

H
Human-centric AI, 161
emphasis on justice, equity, and inclusion, 160
See also Ethical AI guidelines
Humanitarian organization, 129, 131
Human rights, 125–137
Human skills, 1
See also Human thinking
Human thinking, 2

I
Inclusive AI development, 149
addressing inequalities and fostering equity, 146
Information asymmetry, 72
Intellectual property, 108, 111, 112, 115, 122
International (as laws and organization), 134, 136
ISO standards for AI, 155

J
Journalism, 167, 169
Journey, 126, 132, 136

K
Kant, I. (Kantian), 174

L

Laboratório de Políticas Públicas e Internet (LAPIN), 94–101
Latin America, 126, 127, 130
Legal framework, 23, 31
Lithium, 109
Location-based platforms, 70

M

Meta, 108
Methodology
  comparative frequency analysis (CFA), 47, 49
  Large Language Models (LLMs), 47
  LLM methods evaluation, 49–50
  opinion identification and classification, 47–49
  sentiment analysis, 47–49
Mexican, 127, 128
Mexico, 8, 10, 12, 25, 26, 71, 75–76, 127
  See also Mexicans
Migrants, 126–136
Migration, 167
Migration crises, 13
  See also Migration flows
Migration flows, 8, 10
Mozilla Foundation, 93–95
Multi-stakeholder, 87, 89–92, 98–99

N

Napster, 166
National AI strategies, 155
  initiatives in Brazil, Colombia, and Mexico, 155

O

OECD AI principles, 110, 116, 120
Online web-based, 70
OpenAI, 29
Organization of American State (OAS), 122

P

Personal data, 126, 132, 133, 136
Platform economy, 67, 68, 71
Political, 126, 127, 129
Precarization (labor), 22, 24–27
  See also Working conditions
Privacy/private, 126, 128–136
Protect/protection, 126, 129–137
Protocols, 134–136
P2P music interchange, 166
Public perceptions, 42–45, 60
  opinion articles, 45
  policy discussions, 43
Putin, Vladimir, 2

R

Regional declarations on AI
  Santiago Declaration, 159
Regulation, 130–132, 135, 136
  See also Political
Regulation of AI, 147
  global and regional efforts; European Union AI Act, 147; Latin American frameworks, 147
  See also Artificial intelligence governance
Results
  aggregated distribution of sentiments, 52
  public concerns and enthusiasm about AI, 55
Russo-Ukrainian War, 4

S

Security, 126, 128, 131–133, 135
Smart Borders, 128, 129

SMEs, 112
Software, 125, 136
Social media, 22, 23, 25–27, 30, 32, 127, 129, 134, 166, 167, 169, 171, 176
  See also TikTok
Startups, 90, 92, 101
Sustainability, 110, 113, 115, 117, 119, 120
Sustainability of AI, 155, 158, 159
  environmental impact of AI systems, 158, 159

T

Teacher training, 159
  professional development programs for AI integration, 159
Technological companies, 133
Technologies, 126, 128, 129, 131–136
TikTok, 2
Transparency, 86, 88, 92, 93, 96–97, 99, 101
  See also Algorithmic transparency
Transparency in AI systems, 145, 147, 155, 158, 160
  See also Algorithmic bias
The 2020 pandemic, 24

U

UCAV, 166–168, 171, 172
  See also Drones
UNESCO AI ethics framework, 114
UNESCO guidelines on AI, 144, 147
  recommendations for ethical integration in education; Beijing Consensus on AI and Education, 147; Global norms for responsible AI, 147
United States (US), 2, 4–7, 13, 125–132, 135, 136
University initiatives, 156
  institutional approaches in Mexico, Brazil, and Colombia, 156
Uruguay, 70, 71, 76–77
US Supreme Court, 6

V

Venezuela, 25

W

Working conditions, 22
World Intellectual Property Organization (WIPO), 31

The manufacturer's authorised representative in the EU is Springer Nature Customer Service Centre GmbH, Europaplatz 3, 69115 Heidelberg, Germany. If you have any concerns regarding our products, please contact ProductSafety@springernature.com

Printed and bound by CPI Group (UK) Ltd, Croydon, CR0 4YY
30/09/2026
02225595-0001